ANALOG VLSI DESIGN
nMOS and CMOS

PRENTICE HALL
SILICON SYSTEMS ENGINEERING SERIES
Editor: Kamran Eshraghian

HASKARD & MAY	*Analog VLSI Design: nMOS and CMOS*
PUCKNELL & ESHRAGHIAN	*Basic VLSI Design: Systems and Circuits*

ANALOG VLSI DESIGN
nMOS and CMOS

Malcolm R. Haskard
Microelectronics Centre,
South Australian Institute of Technology and
Technical Manager,
Australian Silicon Technology, Adelaide

Ian C. May
Microelectronics Centre,
South Australian Institute of Technology

PRENTICE HALL

New York London Toronto Sydney Tokyo

Prentice Hall, Inc., *Englewood Cliffs, New Jersey*
Prentice Hall of Australia Pty Ltd, *Sydney*
Prentice Hall Canada, Inc., *Toronto*
Prentice Hall Hispanoamericana, S.A., *Mexico*
Prentice Hall of India Private Ltd, *New Delhi*
Prentice Hall International, Inc., *London*
Prentice Hall of Japan, Inc., *Tokyo*
Prentice Hall of Southeast Asia Pty Ltd, *Singapore*
Editora Prentice Hall do Brasil Ltda, *Rio de Janeiro*

Typeset by Monoset Typesetters, Strathpine, Queensland.

Printed and bound in Australia by
Globe Press Pty Ltd, East Brunswick, Victoria.

Cover design by Philip Eldridge

1 2 3 4 5 92 91 90 89 88

ISBN 0-13-032640-2

Library of Congress Cataloging-in-Publication Data

Haskard, M.R. (Malcolm R.), 1936-
Analog VLSI design.

(Silicon systems engineering series)
Bibliography: p.
Includes index.
1. Integrated circuits—Very large scale integration—Design and construction. 2. Linear integrated circuits—Design and construction.
I. May, I.C. (Ian C.), 1959- . II. Title.
III. Series.
TK7874.H392 1987 621.381'73 86-30660
ISBN 0-13-032640-2

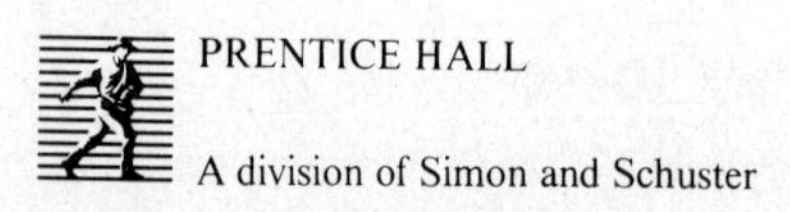

Contents

Foreword

VLSI technology is now maturing rapidly with a current emphasis towards submicron structures. The recognized functional objective is a system-on-a-chip which, to many people, simply means ever more complex digital circuits. However, for many applications, including image and signal processing, the goal of an integrated system implies a *combination* of both digital and analog circuitry in monolithic form. A/D conversion is a key example here, and a candidate for on-chip integration with other digital processing circuitry. Thus the topic of this text *Analog VLSI Design: nMOS and CMOS* is timely, and its appearance is a logical response to the trend towards complete systems integration.

Worldwide there are only a few thousand top VLSI chip designers—only a few hundred of these at most are MOS analog circuit specialists. Predictably, most analog designers specialize in bipolar transistor circuits—the core of MOS analog engineers is small indeed. Because of the recognized skill shortage, there now appears to be support from industry to train more postgraduate student engineers in analog MOS techniques. This book, which provides an excellent introductory treatment of the topic is therefore welcomed as a suitable text for such new courses. Consequently, I have to acknowledge the foresight of Malcolm Haskard and Prentice Hall, Australia, for recognizing the opportunity and producing a volume which in my view meets a vital need.

Professor John Mavor
University of Edinburgh

Preface

In recent years there has been an increasing interest in MOS analog circuits. People have realized that to achieve the full economic gain of silicon technology one has to place the complete system on a chip, including analog as well as digital circuits.

Since its inception in 1970, the Microelectronics Centre has always been involved with bipolar analog circuits, but it was not until 1981 that the first simple MOS analog circuit was designed and fabricated inhouse.

In 1981 a proposal was presented to Dr Craig Mudge, who had recently established the "VLSI Program" Group within CSIRO, that work in the analog multi-project chip area be undertaken. As a result, Malcolm Haskard was invited to join the group for nine months as a consultant in that area. Since that time, the work has continued at the Microelectronics Centre with assistance given by both staff and students, including major contributions by Ian May. The work has concentrated on lambda = 2.5 micron processes, with the minimum transistor dimensions being 5 microns.

At about the time this analog work commenced, the IEEE Press published a volume entitled *Analog MOS Integrated Circuits*, edited by Paul Gray, David Hodges and Robert Brodersen (all then of the University of California, Berkley). This book brought together many of the articles published on MOS analog circuits. We recommend this volume to all who are interested in analog work.

To the many who have encouraged and assisted us we offer our sincere thanks, especially Alan Marriage, Malcolm Macdonald and John Bannigan of the Microelectronics Centre, and to the students who in many cases were the "guinea pigs" for the original lecture series. We believe, however, that there is still much more to be known. We trust that this work will not only provide you with useful information to assist you in your designs, but also encourage you to go on to greater things.

We would also like to express our appreciation to those who provided us with technical information. These include Hugh McDermott of University of Melbourne, Dr Kamran Eshraghian of University of Adelaide, Dr Jim Mason of Queen's University, Kingston, Canada, Chris Meyer and Vladamir Svoboda of the Microelectronics Division of AWA, Sydney, Dr Peter Hicks of UMIST, Manchester, England, and Dr Rob Potter of Austeck Microsystems Pty Ltd, Adelaide.

Finally, we would like to express sincere thanks to the Division of Computing Science, Commonwealth Scientific and Industrial Research Organisation, Austek Microsystems Pty Ltd and the Joint Microelectronic Research Centre at the University of New South Wales who provided us with fabrication services free of charge. While all these organizations are Australian, the foundaries into which they provided free service were mainly overseas ones.

Malcolm R. Haskard and Ian C. May
South Australian Institute of Technology

1 Introduction

1.1 The need for analog circuits

Perhaps one of the greatest revolutions of our age is the advent of the silicon chip. It is now possible to produce a die with over a million transistors on it so that total systems can be produced on a single chip. It is not difficult to imagine the changes that this is introducing to the whole of society. A decade ago a telephone exchange or a powerful computer occupied a room and cost a considerable amount of money to set up and maintain. Today such systems can be accommodated on a single chip at a fraction of the cost, so that private individuals and small companies can exploit the power of these systems.

A part of this revolution is the drastic change in component costs of a system. A decade or so ago active electronic components and the assembly of them onto a printed wiring board constituted the major cost of the system. By integrating the complete system into a single component, these costs are considerably reduced and are often dwarfed by the costs of the subsidiary components such as plugs, sockets, interconnecting cables, switches and the box to house it all in. A further advantage of being able to integrate a total system is the improvement in reliability. As a first order approximation, the reliability of a system is proportional to the number of soldered or made connections. By integrating a system onto a single chip there is a considerable improvement in reliability. Of course the chip itself now becomes the limiting factor in the reliability, but with modern clean rooms and stable fabrication processes, chip reliability is extremely high. Even if a failure should occur, the fault finding and replacement procedure can be minimal, as the system is basically a single chip which is simply replaced.

Having seen the possibility and desirability of including the total system on a chip, the next operation inevitably deals with the problem of how to best implement a system in silicon. Up until two decades ago the majority of systems constructed were of an analog nature. The real world is basically analog and, in general, an analog system requires fewer components than a digital one. The advent of the mini computer in the 1960s started to change this situation. It was economic to explore the possibilities of the alternative digital approach, for the digital computer allowed a degree of decision making and computational ability not possible with an analog system. It was soon realized that most analog functions could be done at least as well and often considerably better, using digital techniques. However, the disadvantage of a digital system was that it required more components, and was therefore more expensive. The rapid advances in silicon technology have now changed this; it is now almost as expensive to make a single transistor as a complete circuit or even a small system on a single die. For these reasons, spurred on by further research in the use and refinement of digital methods, there is an increasing swing towards the use of digital methods. For example, the telephone industry has left behind the conventional analog relay exchange and changed to a pulse code modulation digital system employing computers to process, control and switch signals. The recently announced military high frequency receiver by Rockwell–Collins employs digital signal processing of intermediate frequencies. Digitizing, only possible because of very large scale integration, resulted in a 40 percent reduction in the number of components used, and a 30 percent decrease in cabinet size compared to the usual all analog receiver. In addition the reliability and overall performance has improved with the ability to have bandwidth and other parameters under software control (Iversen, 1984, pp. 45–46).

While designers now prefer to use digital methods, not all analog components can be dispensed with. Because the real world is analog, many of the interfacing circuits must also be analog. In order to place a complete system on a silicon chip, these analog interfacing circuits must therefore also be included. Unfortunately, for many reasons—speed, packing density or simply that it is the only process readily available—many systems must be fabricated using the nMOS process. In numerous countries, the multiproject chip schemes (Pucknell and Eshraghian, 1985, and Mudge et al., 1983) used for teaching and research are predominantly silicon gate nMOS so that analog circuits must be included on what is really a digital circuit process. For this reason it will be assumed in this book that both standard nMOS and CMOS processes will be used for fabrication and that the majority of processing will be undertaken digitally. Analog circuits will only be employed when absolutely necessary and normally for interfacing to the real or physical world (as shown in Figure 1.1).

The question to be answered, then, is: What types of analog circuits are required? Five basic categories are as follows:

1. *Pads.* Analog input and output pads are required so that analog signals can be fed into and taken out from the chip. Signals can be either in voltage or current form. In addition to these pads, we will see later that pads may also be required as test points and for setting precision voltage references.
2. *Amplifiers.* All circuits requiring simple amplification have been grouped under this category. This includes open and controlled gain amplifiers, simple filter circuits and comparators.
3. *Voltage controlled oscillators.* An important circuit block in many systems is a phase locked loop. Part of this block is a voltage controlled oscillator. It differs from a normal oscillator in that an analog input voltage (or current in some instances) determines the frequency of oscillation.
4. *Converters.* These are a very important group of circuits, for they allow analog signals to be converted into digital and vice versa. An important part is the need for a constant voltage (or current) reference.

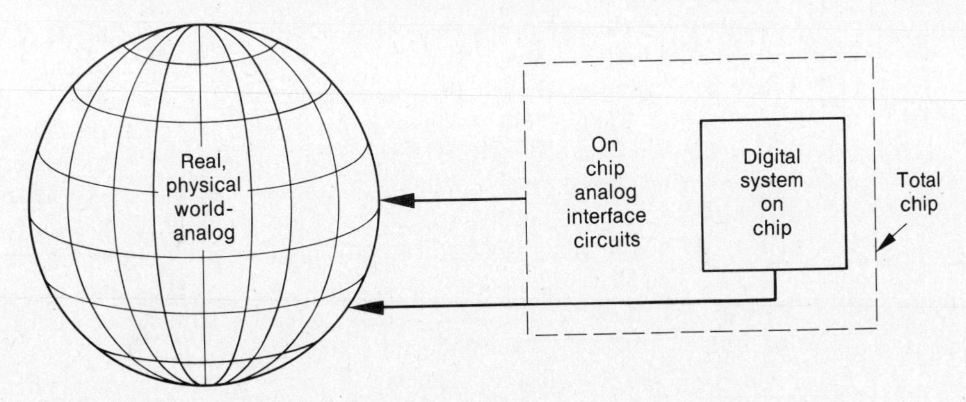

Figure 1.1 The need for analog interface circuits

5. *Sensors*. Silicon as a material displays a number of characteristics that make it useful for making sensors. For example, thermal, magnetic, optical and strain effects can be directly sensed. Other sensors can be made using circuit ideas rather than the silicon properties. Sensors can be further divided into two types, namely those that may need to be replicated and produced as a matrix (e.g. optical, magnetic) and those that in general do not (e.g. temperature, strain).

For each of these analog circuit categories it would be ideal if a library of cells could be designed so that system designers need only call up the appropriate parameterized cell and add it to their layout. The objective of this book is to show how such cells can be designed for standard MOS processes.

1.2 Definitions and terms

1.2.1 Important assumptions

Many companies or educational establishments do not have their own inhouse semiconductor processing facilities to make silicon chips. They therefore must make use of an external semiconductor house, often called a silicon foundry. This can pose problems. For example, the silicon foundry may not provide adequate process and device parameters, and the cost of processing a minimum quantity wafer batch can be prohibitive.

To overcome the latter, Mead and Conway (1980) introduced the multiproject chip scheme, where designs from many organizations are merged into one or more large die and then processed. Typically there may be six to eight designs to a die with four to six die types included on a wafer, giving 30 to 40 designs per fabrication run. In this way costs are shared between many groups. The disadvantage of the scheme, however, is that all designs are in the public domain as the chips received contain all contributors' designs which are accessible not only to the owners of the design, but to all organizations involved. A simple variant of this is the multiproject wafer scheme where each organization has its own die with a design on it, there being several die types to a wafer. Costs are greater, but security can be high as each group receives only its own design.

Before interfacing into any semiconductor foundry, there needs to be an agreed standard interface. Again the concepts proposed by Mead and Conway (1980) have been accepted, namely that the language used to describe the topology of designs by the Caltech Intermediate Form (CIF) and that the designs be according to lambda based rules. Such rules are a simplification of most individual micron (micrometer) based rules supplied by foundries. Lambda (λ) is the smallest unit or increment that a designer can use. Thus for a standard production process, lambda may be 1.5 microns, while for an advanced research line it may be 0.75 micron or less. This is transparent to the designer. While lambda rules do not give the minimum silicon area for design or provide the ultimate in performance extracted from a particular semiconductor manufacturing process, they are very simple and allow the flexibility of changing foundries without having to redo the design. This has advantages for educational establishments and for professional engineers or computer scientists who may like to test circuit or system ideas. When proven, a full commercial design can be undertaken, based on the micron layout rules of the selected semiconductor manufacturer.

With the nMOS process there are two possible methods of interconnecting the n^+ diffusion and polysilicon layers, namely the butting and buried contacts The latter is the more recent, occupying less area, but requiring an additional mask for manufacture. Butting contacts have the advantage that all three conductor layers are available. Throughout this book both are used. With care, the two can be interchanged with no circuit performance degradation.

1.2.2 Layout descriptions

While the Caltech Intermediate Form (CIF) has been selected as the method of describing the topology of a design ready for fabrication, it is not the most convenient for describing layouts in a book. Appendix C does, however, give the description in CIF of an analog test structures chip for determining process parameters. Elsewhere color plates or black and white drawings have been included to illustrate typical design layouts.

1.2.3 The tile concept

In the digital area many complex circuits are constructed from a range of primitive cells sometimes called leaf cells (Newkirk and Mathews, 1983). The same concept can be applied to analog circuits and a very versatile set of primitive cells called tiles have been developed. They are of a standard height and width and normally interconnect by abutment. By using these tiles, complex analog circuits, called a mosaic, can be quickly formed. Chapter 5 and Appendix A provide more information on this concept as well as details of some of the tiles.

1.3 A standard interface to fabrication

In summary, it is essential that both digital and analog circuits including sensors be accommodated on the same chip so that the full impact of silicon technology is realized. Thus, any standard interface to fabrication for analog circuits must be identical to that for the digital. Since on MOS processes it is considerably easier to produce digital circuits, coupled with the fact that many analog operations can now be performed better digitally, the overall philosophy is to employ analog circuits only where absolutely necessary.

Layout of circuits will be to lambda based rules in this book, while the final description of the circuit topology forwarded to a foundry will normally be given in the Caltech Intermediate Form (CIF).

1.4 Exercises

1 Compile a list of systems known to you that, if fabricated on a silicon chip, would require analog interface circuits.

2 For the systems in question 1, produce block diagrams and check that the analog sections can be constructed using the five categories given in section 1.1.

1.5 References

Iversen, W. R. (1984), "Digital HF Radio Breaches Military Solid Analog Wall", *Electronics International*, 28 June, pp. 45–6.

Mead, C. and Conway, L. (1980), *Introduction to VLSI Systems,* Addison Wesley, Reading, Mass.

Mudge, J. C., Clarke, R. J., Paltridge, M. L. and Potter, R. J. (1983), "Results of AUSMPC 5/82", *VLSI Design*, January/February 1983, pp. 52–6.

Newkirk, J. and Matthews, R., *The VLSI Designers Library*, Addison Wesley, Reading, Mass.

Pucknell, D. and Eshraghian, K. (1985), *Basic VLSI Design: Principles and Applications*, Prentice-Hall, Sydney.

2 nMOS: The basic MOS process

2.1 Introduction

Of all the current MOS processes, nMOS is the simplest. Reasons for its popularity include high circuit packing density and the less complex design software needed. Further, it is now a matured process so that most of the silicon gate nMOS processes offered by foundries have very similar characteristics. This is important for second sourcing designs. Excellent texts on digital nMOS design are readily available (Pucknell, Eshraghian, 1985; Mead and Conway, 1980) and this is important for the training and retraining of engineers and computer scientists. Consequently, we will examine this basic process in detail. A description of the two transistor types as well as other passive components available will be given, simple illustrative design examples presented and a description of a typical nMOS fabrication process is provided.

2.2 nMOS transistors

As the name indicates, the nMOS transistor consists in cross section of three layers, Metal (a conductor), Oxide (an insulator) and Semiconductor. The n prefix indicates that the channel through which conduction occurs is of n-type silicon. Figure 2.1 shows diagrammatically the cross section and symbols of such a transistor.

Notice that the transistor is symmetrical and that the drain and source symbols could be reversed. At this stage we will assume the source is connected to the substrate. If a positive potential is applied to the drain with respect to the source, no current flows for there are two back to back diodes (a pn junction) in series. If now a positive potential is also applied to the gate with respect to the source, and by considering the gate thin silicon dioxide insulator with p-type substrate beneath as a capacitor, the gate voltage, being positive, induces negative charges at the surface of the substrate under the gate, depleting the p-type hole carriers (Figure 2.2). At some potential on the gate, the induced charge is sufficient to cause inversion, so that a thin n-channel region is formed at the surface of the substrate under the gate,

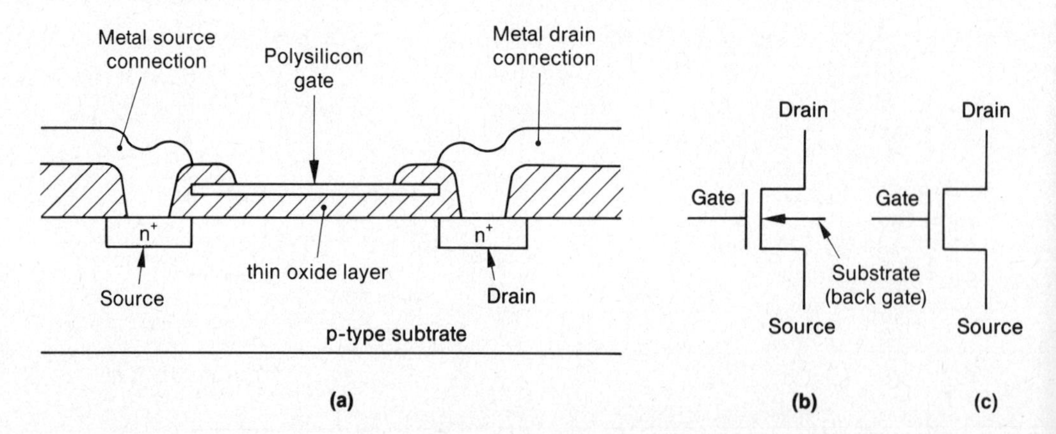

Figure 2.1 nMOS transistor transistor construction and symbols

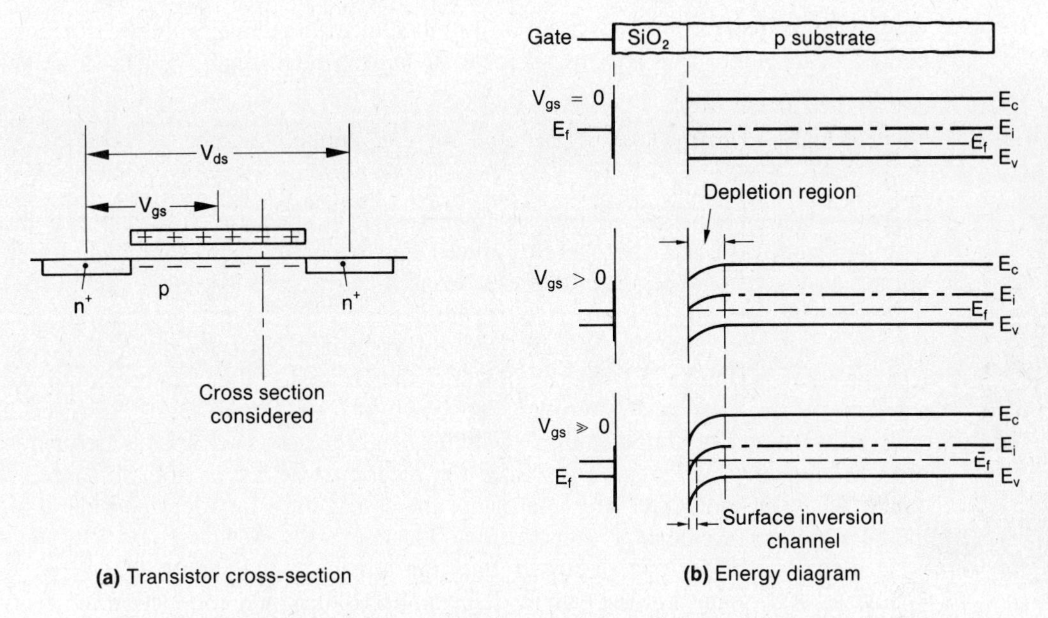

Figure 2.2 Transistor operation explained in terms of energy diagrams for the gate, oxide, substrate, and MOS capacitor

allowing conduction to occur between the source and drain. The gate source voltage at which conduction can just occur is called the threshold voltage. For small drain source voltages the channel appears approximately as a linear resistance because there is a potential gradient along the channel caused by the drain positive potential, for larger drain source potentials nonlinearities occur.

At some point, an increasing drain voltage becomes dominant and the channel current becomes pinched or saturated as shown in Figure 2.3. Thus the

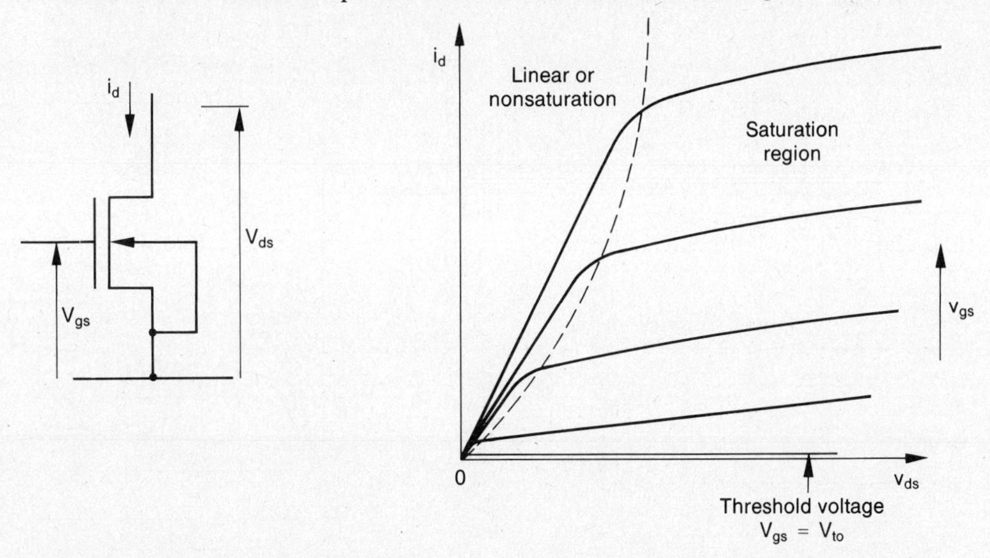

Figure 2.3 Static characteristics

characteristics (correctly called drain or output static characteristics) of the transistor can be divided into two regions, a linear or nonsaturated region and a saturated region.

In the discussion this far, the substrate and source have been connected together. If this is not the case, then any potential between the substrate and source (normally, the substrate is made negative with respect to the source to hold the source and drain substrate diodes reverse biased) will affect the voltage across the gate substrate capacity, the charge induced and hence the gate source threshold voltage at which conduction commences. Consequently, the substrate is often called the back gate as it also has a controlling influence on the drain source current and threshold voltage. Thus the symbol for the transistor is correctly given as a four terminal device (Figure 2.1(b)). In practice, the source is often connected to the substrate so the back gate is omitted (as shown in (c) of Figure 2.1). However, a word of warning: for simplification of circuit diagrams, the back gate is frequently omitted altogether as it is obvious from the diagram whether or not the source is connected to the substrate. Any analysis of the circuit must take into consideration the back gate voltage effect.

The transistor type discussed thus far is called an enhancement type. A voltage must be applied to the gate for the channel to form and conduction to occur. It is possible during manufacture of the device to implant a controlled dose of the correct impurity type ions to form at the substrate surface a small n-type channel. Consequently, for zero or no gate voltage, conduction will occur between the source and drain as there is simply an n channel present. A positive gate voltage increases this current while a negative gate voltage with respect to the source decreases it. Should a large enough negative gate voltage (typically less than −3 volts) be applied, the induced positive charges can prevent conduction occurring altogether and the device becomes cut off. This transistor type is called a depletion nMOS transistor and it has a negative threshold voltage. The back gate voltage also influences the threshold voltage and drain source current. A cross section of a depletion nMOS transistor together with its symbol and static characteristic is shown in Figure 2.4.

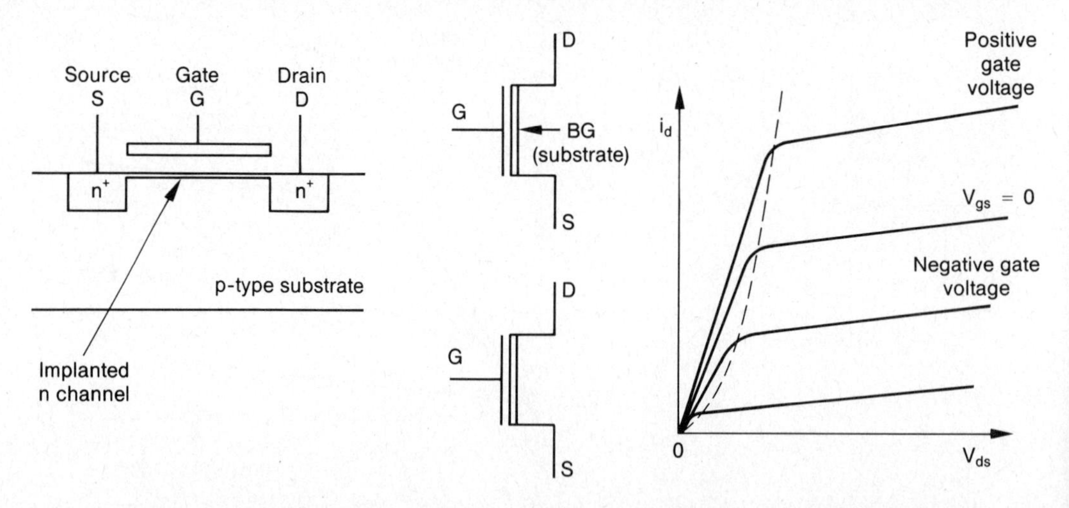

Figure 2.4 Construction, symbols, and static characteristics of a depletion type nMOS transistor

Thus the nMOS process provides the designer with two transistor types: enhancement and depletion. To design circuits using these devices both graphical and analytical methods can be employed, although for computer analysis the latter must be used. To do this, a mathematical model is required of the transistor.

2.3 Models to analyze transistor circuits

To produce a circuit using nMOS transistors, designers must have some model to give them an understanding of how the devices will behave in the circuit. Models which accurately predict the characteristics of the transistor can be extremely complex and not at all suitable for manual calculation. On the other hand, if a computer is employed to handle the complex model, circuit design can become a hit or miss method if designers do not have a simple model to determine the best approach to a satisfactory design. Consequently, in this section a simple model will be developed that can be used as an aid to understanding circuit operation and therefore configuration. The detailed calculations, once the circuit configuration has been determined, can be performed using a computer analog simulation program such as SPICE. Thus the design approach recommended is shown in simplified form in Figure 2.5. Notice that it is an iterative process in which the idea is first tested using a simplified model to establish whether or not it is correct.

To develop a simple model, consider the enhancement nMOS transistor shown in Figure 2.6. We will continue to assume that the source and substrate are connected. A voltage on the gate (V_{gs}) induces charge at the substrate surface and, if the voltage is large enough, strong inversion occurs resulting in a large drain source current. It should be noted that for gate source voltages, less than the threshold voltage ($V_{gs} < V_{to}$) very small currents do flow, for the MOS transistor can behave like a lateral npn transistor, the source being the emitter, drain the collector and substrate the base of width L, which is the channel length. See Figure 2.6(c). We will not consider this region of weak inversion any further as silicon fabricators tend not to control the characteristics of devices in this area.

For the normal or strong inversion region of operation, the drain current flowing can be calculated as follows. The operation is that charge on the gate controls the movement of charge between source and drain in the channel under the gate. Thus,

$$I_d = \frac{\text{Charge induced in channel}}{\text{Transit time}} = \frac{Q}{\tau}$$

where

$$\tau = \frac{\text{distance electrons have to move}}{\text{average velocity}} = \frac{L}{v}$$

$$= \frac{L}{\mu E} \qquad (2.1)$$

$$= \frac{L}{\mu} \times \frac{L}{V_{ds}} = \frac{L^2}{\mu V_{ds}} \qquad (2.2)$$

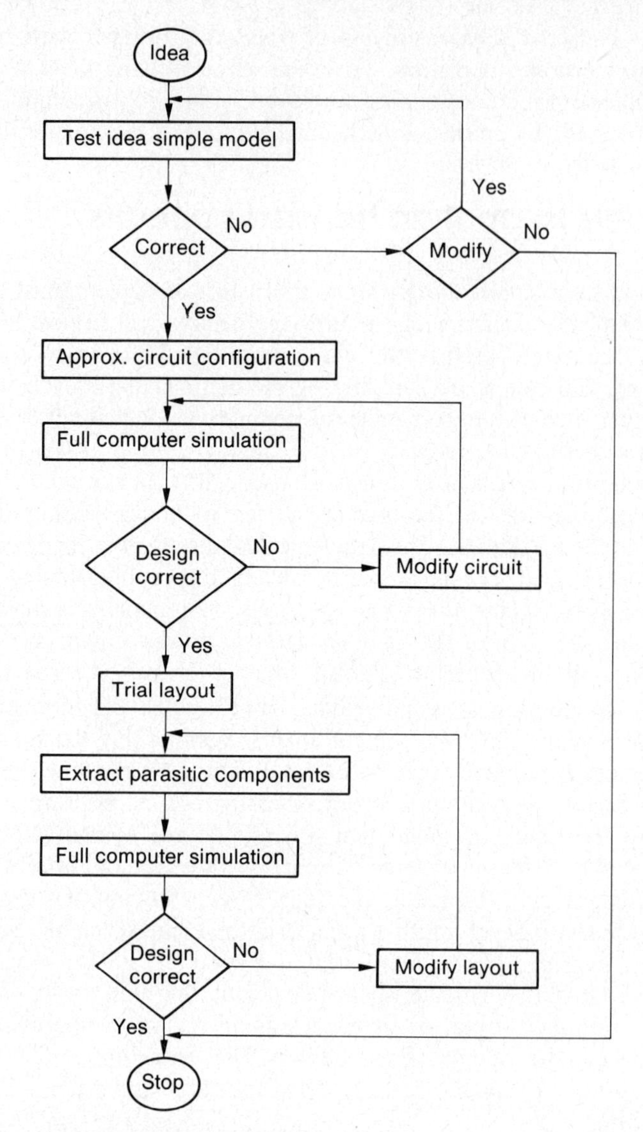

Figure 2.5 The design approach

E being the electric field along the gate due to the drain source voltage,

$$E = \frac{V_{ds}}{L}$$

and μ the mobility of the charge carriers under the influence of the field.

The amount of charge Q in the channel is given by the product of the gate capacity and the gate voltage in excess of the threshold voltage (called the excess gate voltage). For a parallel plate capacitor, the gate capacity is,

$$C_g = \frac{\epsilon A}{D} = \frac{\epsilon WL}{D} \tag{2.3}$$

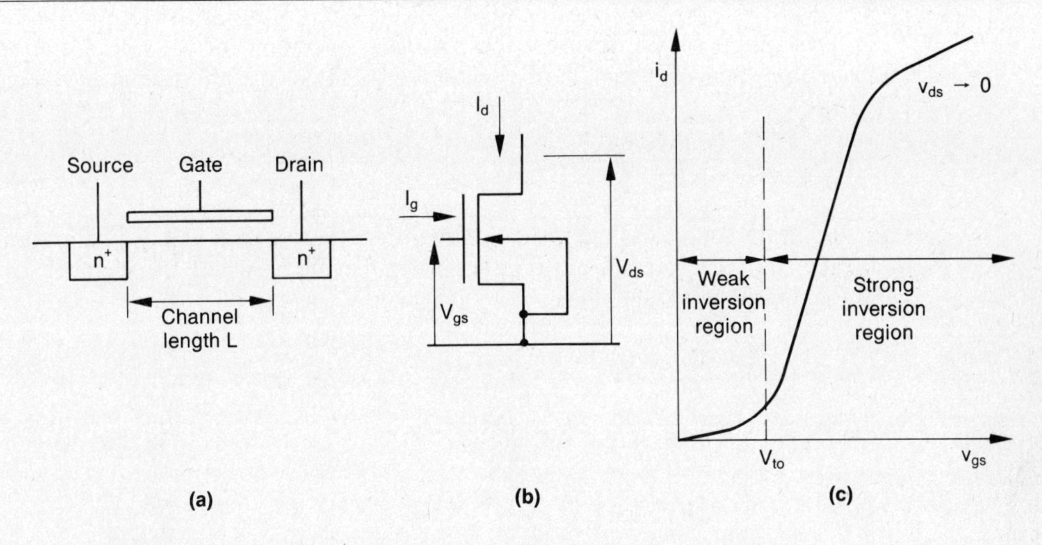

Figure 2.6 Operation of an nMOS enhancement type transistor

where

$$\begin{aligned} \epsilon &= \text{permittivity of the oxide dielectric} \\ A &= \text{gate or channel surface area} \\ &= \text{width W} \times \text{length L} \\ D &= \text{the oxide thickness} \end{aligned}$$

Assuming a linear field along the channel, the average channel voltage under the gate is half the drain source voltage, that is $V_{ds}/2$. The excess voltage across the gate capacitor is, therefore, given by,

$$V_{cex} = V_{gs} - V_{to} - V_{ds}/2 \tag{2.4}$$

Thus the channel charge in transit,

$$\begin{aligned} Q &= C_g V_{cex} \\ &= \frac{\epsilon WL}{D}(V_{gs} - V_{to} - V_{ds}/2) \end{aligned} \tag{2.5}$$

Combining equations 2.1, 2.2 and 2.5 we obtain,

$$I_d = \frac{\mu\epsilon}{D}\frac{W}{L}(V_{gs} - V_{to} - V_{ds}/2)V_{ds} \tag{2.6}$$

$$= \mu C_{ox}\frac{W}{L}V_{ds}(V_{gs} - V_{to} - V_{ds}/2) \tag{2.7}$$

where $C_{ox} = \epsilon/D$ is the capacity per unit area of the gate.

This simple model describes the transistor operation for the nonsaturation region or for when the drain source voltage is less than the effective gate voltage (V_E). Thus,

$$V_{ds} < V_{gs} - V_{to} = V_E \tag{2.8}$$

When the drain source voltage rises to equal the effective gate voltage, saturation commences, for the gate substrate potential at the drain end of the channel just equals the threshold voltage. Thus,

$$V_{ds} = V_{gs} - V_{to} = V_E \tag{2.9}$$

Combining equation 2.9 with 2.6 we obtain,

$$I_d = \frac{\mu\epsilon}{D}\frac{W}{L}(V_{gs} - V_{to})(V_{gs} - V_{to} - (V_{gs} - V_{to})/2)$$

$$= \frac{\mu\epsilon}{2D}\frac{W}{L}(V_{gs} - V_{to})^2 \tag{2.10}$$

or

$$I_d = \frac{\mu C_{ox}}{2}\frac{W}{L}(V_{gs} - V_{to})^2 \tag{2.11}$$

For drain source potentials greater than that given by equation 2.9, the excess voltage is dropped across the depletion region that extends out along the channel from the drain. Saturation is occurring and the drain current flowing is still given by equation 2.11.

For the nonsaturation or linear region,

$$V_{ds} < V_{gs} - V_{to}$$

$$I_d = \mu C_{ox}\frac{W}{L}V_{ds}(V_{gs} - V_{to} - V_{ds}/2) \tag{2.12}$$

and for the saturation region,

$$V_{ds} \geqslant V_{gs} - V_{to}$$

$$I_d = \frac{\mu C_{ox}}{2}\frac{W}{L}(V_{gs} - V_{to})^2 \tag{2.13}$$

These equations highlight a number of issues about our transistor:

1. The magnitude of the drain current can be scaled simply by changing the aspect ratio of the device, that is, the length to width ratio. This is illustrated in Figure 2.7. Transistor (a) has twice the drain current of (b) for the same bias voltages.

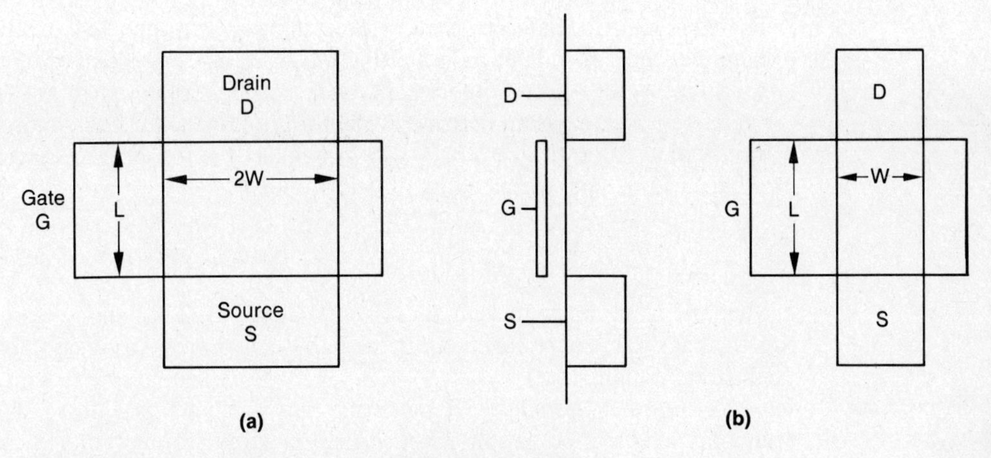

Figure 2.7 Two transistors of various aspect ratio

2. For the nonsaturation region, when the drain source voltage tends to zero, the squared term becomes insignificant and can be ignored. Thus,

$$I_d = \mu C_{ox} \frac{W}{L}(V_{gs} - V_{to})V_{ds}$$

$$\frac{V_{ds}}{I_d} = \frac{L}{\mu C_{ox} W(V_{gs} - V_{to})} \qquad \textbf{(2.14)}$$

The transistor behaves as a linear resistor (as was mentioned in section 2.2), the value of the resistor being determined by the device aspect ratio and gate voltage.

3. For this simple model there are three process parameters of importance: μ the carrier mobility, C_{ox} the gate capacity per unit area, and V_{to} the threshold voltage. In many instances, the first two are combined to form a single process parameter, the intrinsic transconductance denoted by β or K. Thus,

$$\beta \text{ or } K = \mu C_{ox} \qquad \textbf{(2.15)}$$

Since equation 2.13 contains a $\mu C_{ox}/2$ term, an alternative process constant can be used. Thus,

$$K' = \frac{\mu C_{ox}}{2} \qquad \textbf{(2.16)}$$

Care must be taken to check what constant a model may use, for K and K′ can be confused.

4. Although the simple model was developed for an enhancement transistor, exactly the same reasoning can be applied to a depletion transistor resulting in identical equations. Parameter values, however, differ. Table 2.1 shows typical values for transistors.

5. These simple models do not tell us everything about our transistors. For example, they do not include any back gate effects previously discussed. As a first order approximation, the back gate modulation voltage effect can be represented as an increase in the threshold voltage. Designating the threshold voltage with zero back gate voltage as V_{to}, the effective threshold voltage can be approximated by (Tsividis, 1978)

$$V_t = V_{to} + \gamma(\sqrt{-V_{bs} + 2\phi_F} - \sqrt{2\phi_F} \qquad \textbf{(2.17)}$$

where ϕ_F is the Fermi potential (approximately 0.3–0.35) and γ (gamma) is the back gate effect constant for the particular process (a function of the substrate doping level).

Another nonideal characteristic not included is known as the channel length modulation effect. Since the drain source voltage causes the depletion region around the drain channel junction to vary, the channel length L (Figure 2.6) decreases as V_{ds} increases. Consequently in the saturation region I_d is not constant but increases with increasing drain source voltage. A further parameter called lambda can be introduced to allow for this effect. Both the drain current equations 2.12 and 2.13 are modified by a value $(1 + \lambda V_{ds})$. Since λ is small (0.005 to 0.07) an alternative form of $1/(1 - \lambda V_{ds})$ is often used. Thus equations 2.12 and 2.13 become for,

$$V_{ds} < V_{gs} - V_{to}$$

$$I_d = \mu C_{ox}\frac{W}{L}V_{ds}\left(V_{gs} - V_{to} - \frac{V_{ds}}{2}\right)(1 + \lambda V_{ds})$$

or $\qquad$ **(2.18)**

$$= \mu C_{ox}\frac{W}{L}\frac{V_{ds}}{(1 - \lambda V_{ds})}\left(V_{gs} - V_{to} - \frac{V_{ds}}{2}\right)$$

and for,

$$V_{ds} \geqslant V_{gs} - V_{to}$$

$$I_d = \frac{\mu C_{ox}}{2}\frac{W}{L}(V_{gs} - V_{to})^2(1 + \lambda V_{ds})$$

or $\qquad$ **(2.19)**

$$= \frac{\mu C_{ox}}{2}\frac{W}{L}\frac{(V_{gs} - V_{to})^2}{(1 - \lambda V_{ds})}$$

6. The model derived is a static or DC model and does not contain any capacitors. From the discussion thus far, the major capacity associated with the device is the gate capacitor. However, the source and drains have junction capacitors to substrate and there are overlap capacitors from source and drain to the gate. For simple hand calculations, the gate capacity is probably all that needs to be considered but, when using

Table 2.1 Typical process parameters for enhancement and depletion nMOS transistors

Parameter	*Enhancement nMOST*	*Depletion nMOST*
Threshold voltage (volts)	$V_{to} = 0.7$	$V_{to} = -3.5$
Intrinsic transconductance ($\mu A/volt^2$)	$K = 25$	$K = 20$

computer simulation, all capacities should be included. We will discuss capacitors in more detail in section 2.6.

Before using these simple models to examine a circuit, three further matters must be mentioned. First, Chapter 9 discusses in more detail models used in computer simulation and a method of determining the important parameters for these models. Figure 2.8 shows the SPICE model used to simulate an nMOS transistor. It is a nonlinear model including the drain and source diodes, junction and gate capacities as well as the intrinsic drain and source resistances. Second, the values for the parameters—even those included in the simple model discussed—can vary considerably from process to process. Table 2.2 illustrates this point giving figures for several foundries. With multiproject chips, designs may be sent to any one of a number of manufacturers so that successful designs must be capable of operating over a very large spread of parameter values.

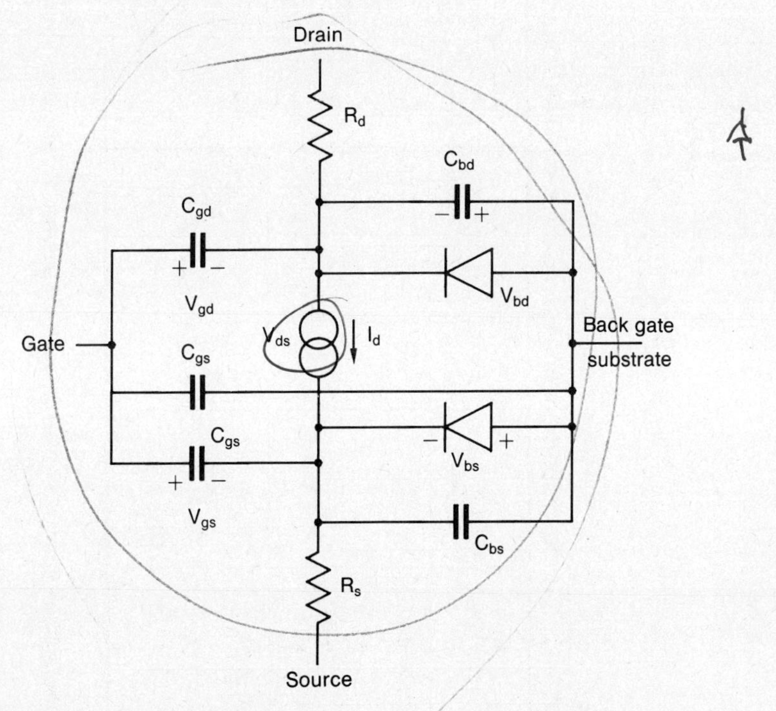

Figure 2.8 SPICE model of a MOS transistor

Table 2.2 Typical process parameter for a number of nMOS foundries

Transistor type	*Parameter*	*Manufacturer 1*			*Manufacturer 2*			*Manufacturer 3*			*Manufacturer 4*			*Unit*
		Min.	Typ.	Max.	Min.	Typ.	Max.	Min.	Typ.	Max.	Min.	Typ.	Max.	
Enhancement	V_{to}	0.65		1.15	0.6		1.0	0.8		1.2	0.8	1.1	1.4	V
	k	25		35	20		24		15		20	25	35	$\mu A/V$
	λ		0.06			0.07						0.02		V^{-1}
	Back gate γ	0.22		0.42	0.7		0.9		0.4		0.35	1.0	1.2	$V^{\frac{1}{2}}$
	x_j		0.5			1.75			1.25			1.0		μm
Depletion	V_{to}	−3.5		−2.5	−4.0		−3.0	−4.3		−3.7	−5.0	−3.5	−2.5	V
	k	30		35		25			6.5		15	22	30	$\mu A/V$
	λ		0.04			0.04						0.02		V^{-1}
	Back gate γ		0.33			0.2			0.6		0.35	1.0	1.2	$V^{\frac{1}{2}}$
	x_j		0.5			1.75			1.25			1.0		μm

Note:
These figures are measured for zero substrate bias, apart from Manufacturer 3 where a reverse bias of −2.5 volts was applied.

The third point is the derivation of a simple low frequency model for small signal analysis. This is important when commencing the design of an amplifier.

If we consider the SPICE model of Figure 2.8, then for a single bias point all the nonlinear diodes are reversed biased and can be ignored, while for low frequency operation all capacitors can be omitted. The resulting model, as shown in Figure 2.9, consists of a dependent current generator and parallel channel conductance.

Values for the two parameters can be computed from equations 2.18 and 2.19. Consider the saturation region by way of example. Thus,

$$\begin{aligned} g_m &= \left.\frac{dI_d}{dV_{gs}}\right|_{V_{ds} = \text{const.}} \\ &= \frac{KW}{L}(1 + \lambda V_{ds})(V_{gs} - V_{to}) \\ &= \sqrt{\frac{2KWI_d}{L}(1 + \lambda V_{ds})} \end{aligned} \tag{2.20}$$

while

$$\begin{aligned} g_d &= \left.\frac{dI_d}{dV_{ds}}\right|_{V_{gs} = \text{const.}} \\ &= \frac{KW\lambda}{2L}(V_{gs} - V_{to})^2 \\ &= \lambda I_d/(1 + \lambda V_{ds}) \end{aligned} \tag{2.21}$$

This simple model immediately tells us that:

1. the maximum possible stage gain is g_m/g_d;
2. that since g_m/g_d is proportional to $1/\sqrt{I_d}$ (from equations 2.20 and 2.21), the gain decreases with increasing drain current.

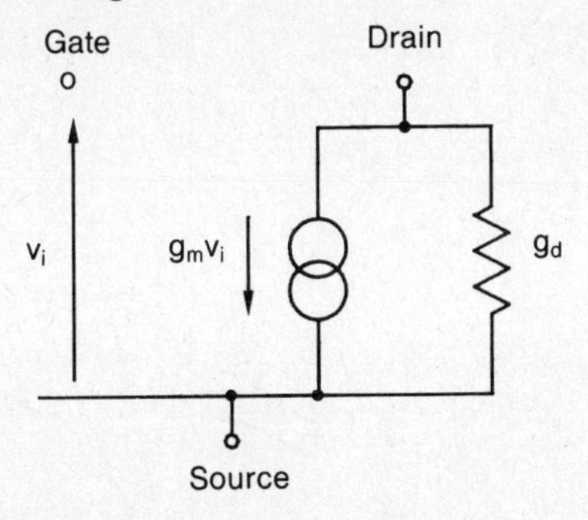

Figure 2.9 Simple small signal model for a MOS transistor

2.4 A simple circuit design

To test our simple analytical model, we can apply it to a standard digital inverter circuit, as given in Figure 2.10. The characteristic curve of $V_{gs} = 0$ for M2 can be drawn as a load line on the characteristics of M1, as shown in Figure 2.10. It should be noted that this is a simplistic approach to illustrate concepts, for it ignores the back gate voltage effects, the source of M2 not being at the substrate potential. Following the design philosophy outlined in Figure 2.5, a detailed computer simulation using a complete model must always follow.

Examining the OFF operating point, we can recognize from the two drain characteristics that M1 is in the saturation region while M2 is in the nonsaturation region. Since there is no external load current, the two transistor drain currents are equal and we therefore have (using equations 2.12 and 2.13),

$$I_d = \frac{\mu C_{ox}}{2}\frac{W_1}{L_1}(V_i - V_t)^2$$
$$= \mu C_{ox}\frac{W_2}{L_2}\overline{V}_o\left(-V_{t2} - \frac{\overline{V}_o}{2}\right)$$

where

$$\overline{V}_o = V_{dd} - V_o$$

Expanding and rearranging we obtain,

$$\overline{V}_o^2 + 2\overline{V}_{t2}V_o + \frac{W_1L_2}{L_1W_2}(V_i - V_{t1})^2 = 0$$

and on solving,

$$\overline{V} = V_{dd} - V_o = -V_{t2} - \left[V_{t2} - \frac{W_1L_2}{L_1W_2}(V_i - V_{t1})^2\right]^{\frac{1}{2}} \qquad \textbf{(2.22)}$$

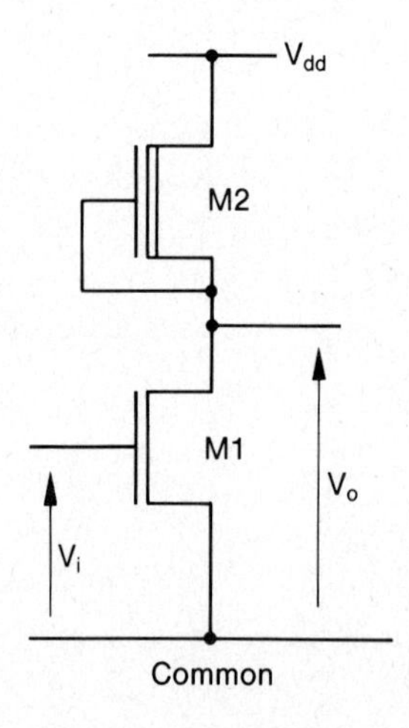

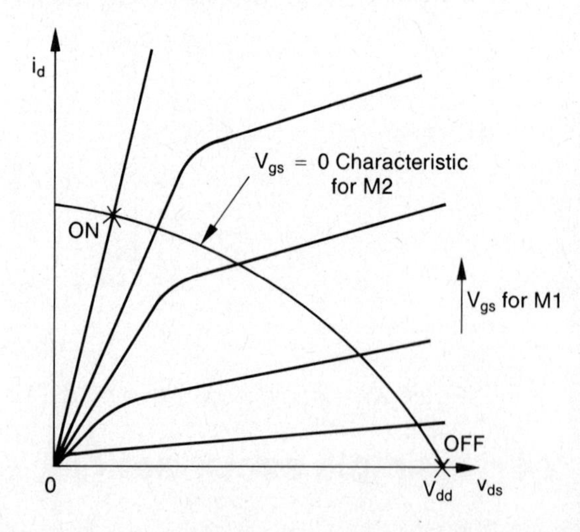

Figure 2.10 Standard digital inverter

If we now consider the ON region of operation, M1 is now in the nonsaturation region while M2 is in the saturation region. Thus using the same reasoning, we can write,

$$I_d = \mu C_{ox} \frac{W_1}{L_1}(V_o)\left(V_1 - V_{t1} - \frac{V_o}{2}\right)$$
$$= \frac{\mu C_{ox}}{2} \frac{W_2}{L_2}(-V_{t2})^2$$

or

$$V_o^2 - 2(V_i - V_{t1})V_o + \frac{W_2 L_1}{L_2 W_1}(-V_{t2})^2 = 0$$

Solving, we obtain,

$$V_o = (V_1 - V_{t1}) - \left[(V_1 - V_{t1})^2 - \frac{W_2 L_1}{L_2 W_1}(-V_{t2})^2\right]^{\frac{1}{2}} \quad \textbf{(2.23)}$$

Equations 2.22 and 2.23 can be plotted to give the transfer characteristic of the inverter. We will assume $V_{t1} = 0.7$ volts, $V_{t2} = -3$ volts, $V_{dd} = 5$ volts, $W_1/L_1 = 3$ and $W_2/L_2 = \frac{1}{3}$. The result is plotted in Figure 2.11.

An interesting point can be seen from this transfer characteristic. When switching from high to low (output), a sudden transition occurs when the input voltage is of the order of 1.5 volts. This we can consider as the inverter threshold voltage. Further, in this region the steep transfer characteristic indicates that the inverter gain is large. For example, from equation 2.22, when the input signal changes from 1.5 to 1.7 volts, the output changes from 3.8 to 1.0 volts, that is, an incremental gain of −14. Thus, if we could bias the inverter into this region, we

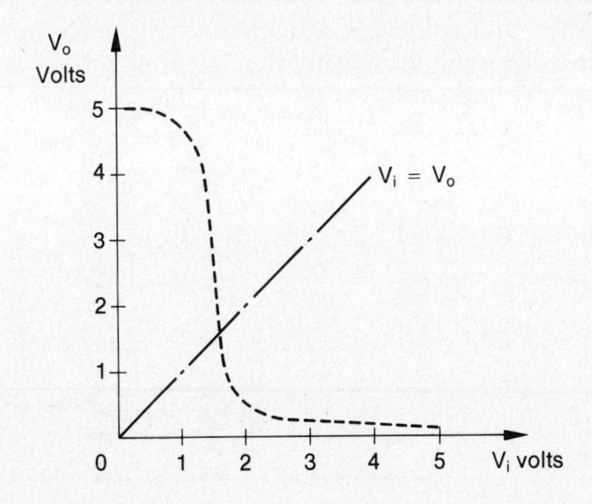

Figure 2.11 Transfer characteristics of the inverter

could use it as a linear amplifier with a gain of −14, the negative sign indicating a phase inversion.

Adding a series pass transistor to this circuit converts the circuit into half of a shift register, as shown in Figure 2.12. The input signal V_i is clocked through M3 onto the gate of M1 where the gate capacity stores it. For V_{i1} less than the inverter threshold voltage, the inverter will switch off; for V_{i1} greater than the threshold voltage, it will turn on. However, what is the relationship between voltage V_{i1} and V_i? Is any voltage lost across M3?

We can check this out by using our simple model. The input voltage V_i is connected to what we may arbitrarily call the drain of transistor M3. To turn the transistor M3 on, the gate source voltage must exceed the threshold voltage V_{t3}. Since we require V_{i1} to approximately equal V_i, then M3 must operate in the nonsaturation region, that is,

$$I_d = C_{ox}\frac{W_3}{L_3}V_{ds}\left(V_{gs} - V_{t2} - \frac{V_{ds}}{2}\right)$$

where

$$V_{gs} = V_\phi - V_{i1}$$

and

$$V_{ds} = V_i - V_{i1}$$

The magnitude of the current that flows through M3 to charge up the gate capacity of M1 is therefore dependent on the voltage $V_\phi - V_{i1} - V_{t3}$. In other words, to maintain a given current flow there must be a corresponding gate source voltage to support it. Thus V_{i1} must always be less than $V_\phi - V_{t3}$ and therefore for V_{i1} to rise to the drain voltage V_i, V_i must be less than $V_\phi - V_{t3}$.

Using the simple model, we have established the operation of two simple circuits. The circuit ideas must be checked further using more precise models, such as that shown in Figure 2.8 and parameter spreads of Table 2.2. This can be achieved by using a suitable analog simulation program such as SPICE.

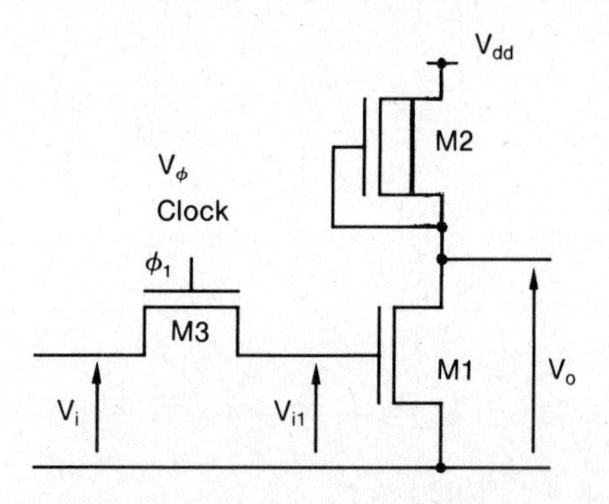

Figure 2.12 Modified inverter to form half a shift register

2.5 The nMOS fabrication process

This process is called a planar process because it is concerned with several layers, each layer having two dimensions. The inverter discussed in section 2.4 is laid out as a drawing of six layers in a similar way to that of a printed wiring board. To transfer the pattern of each layer to the silicon wafer, a photolithographic process is used. An ultraviolet light sensitive photoresist coating is used so that selected portions of the resist are removed, exposing parts of the wafer below for processing. Consider the example shown in Figure 2.13. A layer of silicon dioxide (or possibly silicon nitride) covers the silicon surface.

Photoresist is applied in a uniform layer, then baked and exposed to ultraviolet light through a mask containing the pattern to be transferred. In this case of a positive resist, the exposed resist polymerizes and becomes soluble in the developer, so that on washing it is removed, exposing the silicon dioxide beneath. This can then be etched in hydrofluoric acid; thus the exposed silicon dioxide is etched away to reveal the bare silicon. Finally the remaining photoresist can be stripped off, leaving the pattern that was on the mask transferred to the oxide. This process of pattern transferral is often called *patterning*.

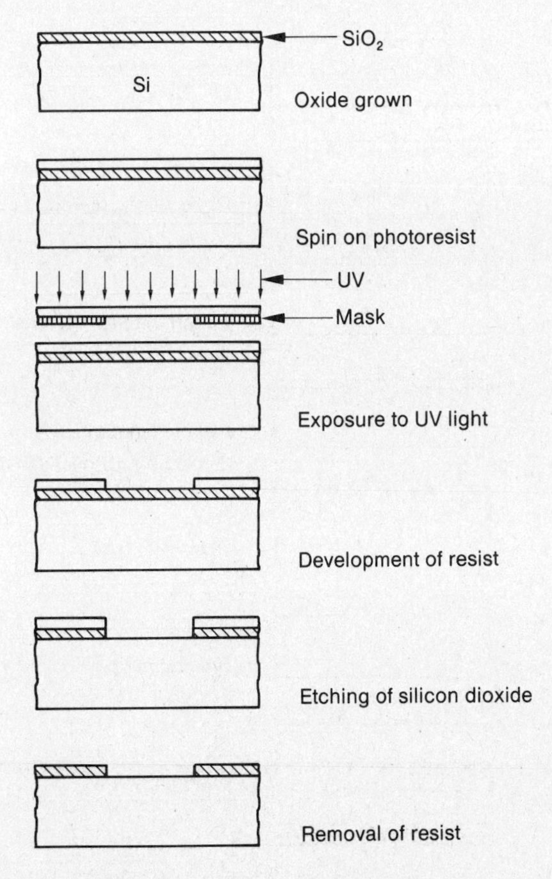

Figure 2.13 The photolithographic process (patterning)

The six layers that are required to produce six masks are:

1. diffusion masks for N^+ region (color green);
2. polysilicon area for gates (color red);
3. depletion implant to make depletion transistors (color yellow);
4. contact holes through protecting oxide layers (color black or sometimes white);
5. metal mask for aluminium interconnection (color blue);
6. overglaze mask to open up holes in the protecting overglaze to allow access for bonding and probing (color orange or brown).

The colors given in brackets are the standard colors used on many computer plots to identify each layer (Pucknell and Eshraghian, 1985; Mead and Conway, 1980).

The processing steps are given in Figure 2.14, starting with a p-type silicon wafer of resistivity typically 20 to 30 ohm cm. A silicon nitride mask is grown over the surface by chemical vapour deposition methods. This is then patterned using the

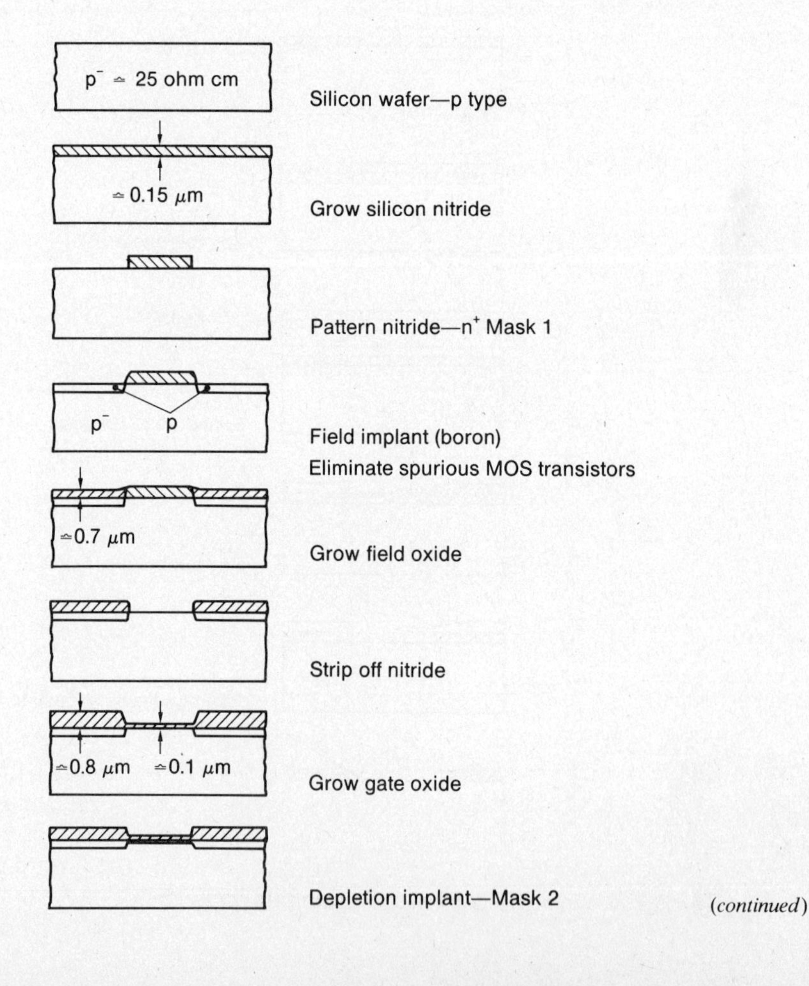

(continued)

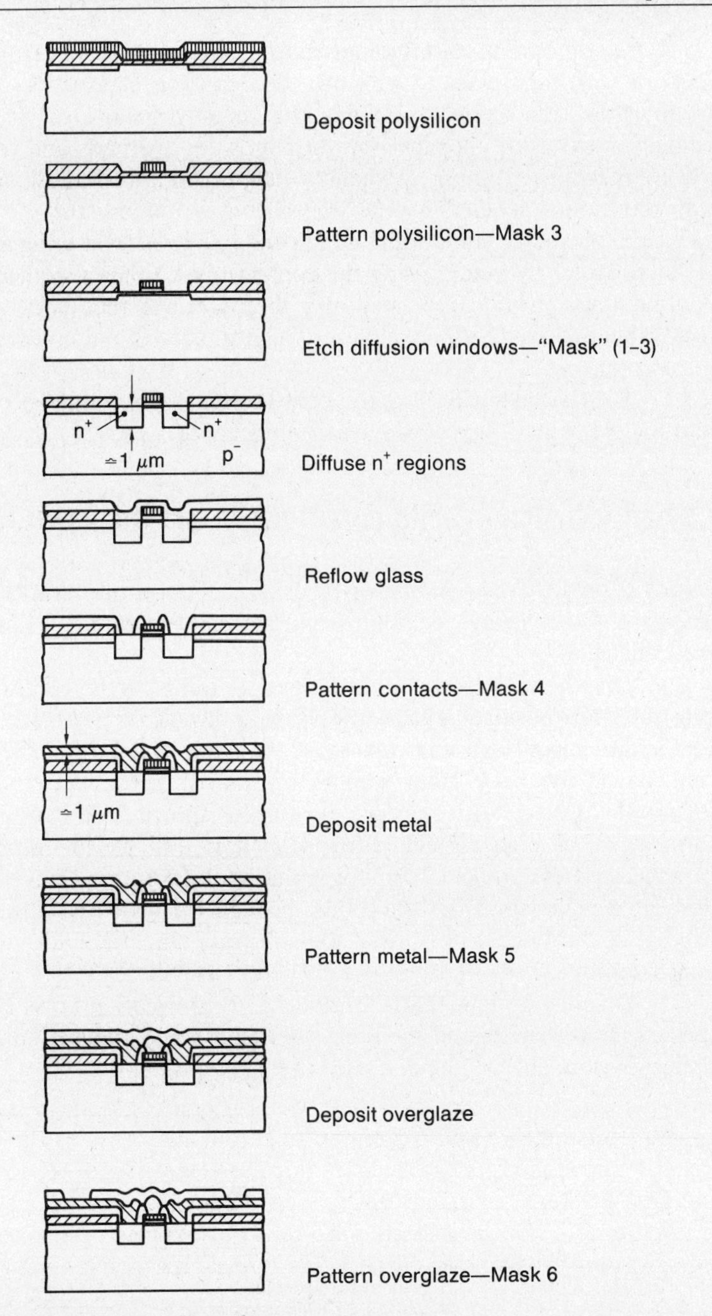

Figure 2.14 Steps in the nMOS fabrication process

diffision (green) mask. A boron ion implant is carried out so that in all areas except where the diffusion (transistors) is to occur, a p layer is implanted. This is to stop spurious MOS transistors occurring where metal conducting tracks crossing diffusions can cause conduction between those diffusions, the thick field oxide acting as the gate oxide. The next step is to grow the field silicon dioxide layer thermally. The nitride layer is now stripped off and the thin gate oxide is grown thermally.

At this point, the implant mask (yellow) is used to determine where the implant is to take place to produce the depletion transistors. The implant occurs through the thin oxide. Following this, a polysilicon layer is deposited over the whole surface using chemical vapour deposition methods and then patterned using the polysilicon (red) mask. A shallow etch, called a wash, now occurs to remove the thin oxide, not covered by the polysilicon, to expose the silicon surface for the diffusion of source and drains. Following this, a layer of glass is spun on and reflowed over the wafer. Using the contact mask (black), contact holes are etched. Aluminium is vacuum deposited over the wafer and then patterned using the metal mask (blue). Finally, the overglaze is deposited and patterned using the overglaze mask (orange).

The finished wafer is then probed to see if it is within specifications and sawn into the individual chips which are then packaged.

2.6 Other components available on the nMOS process

Thus far only the two transistor types available on the nMOS process have been discussed. Other simpler components are available and these include diodes, resistors and capacitors.

The diode is formed using the n^+ diffusion in conjunction with the p-substrate. For positive voltages the diode is always reverse biased. If the diode is to be forward biased extreme care must be taken, for the resulting substrate current can change the back gate voltages of transistors, so that the circuits may not function correctly. An example of this is shown in Figure 2.15. Because the substrate is of high resistance material, it is difficult to make a good substrate contact, so there are usually many point contacts rather than a total area contact. Any current flowing from the diode through the high resistance substrate material will cause a voltage drop so that the substrate potential under the transistors is no longer zero.

Despite the fact that the diode of the nMOS process should normally be used in the reverse biased mode, it does have some applications, particularly as an optical sensor. This will be discussed in Chapter 8.

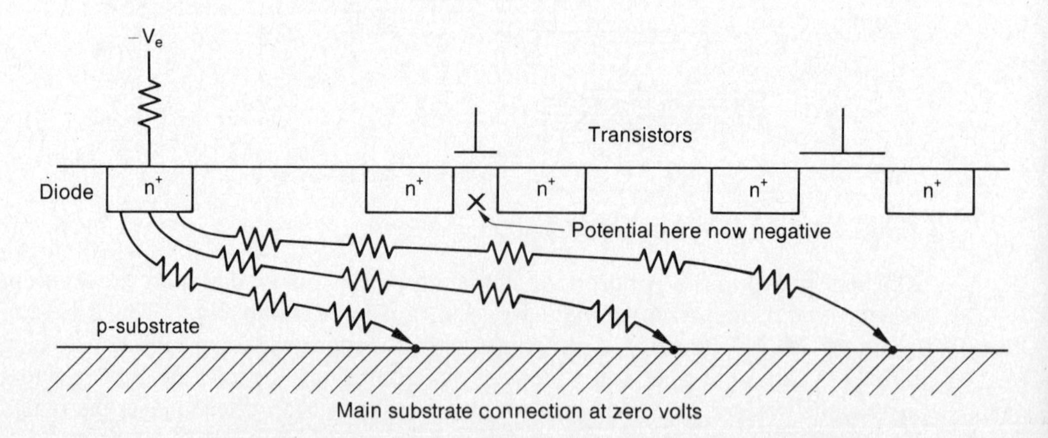

Figure 2.15 Effect of substrate currents

Three types of resistors are available: the n^+ diffusion layer, polysilicon layer and MOS transistor (normally of depletion type), previously discussed. Table 2.3 gives typical parameters for these types of resistors. The resistance value is expressed as a sheet resistance R_s, that is, a constant thickness (or depth) is assumed so that only the length to width ratio is required to define the resistance value. Thus,

$$R = \frac{\rho L}{A} = \frac{\rho L}{t.W}$$

$$= R_s \frac{L}{W} \qquad \textbf{(2.24)}$$

where ρ is the specific resistance of the material. Thus,

A = the resistor cross-sectional area of thickness t and width W
L = the resistor length
R_s = the sheet resistance.

Resistors of the same aspect ratio L/W have the same value of resistance but larger resistors can dissipate more power since the total current flowing can be greater for the same current density.

Table 2.3 Resistor parameters

	Resistor type						
	Diffused		Polysilicon		Depletion transistor		
Parameter	Manuf. 1	Manuf. 2	Manuf. 1	Manuf. 2	Manuf. 1	Manuf. 2	*Units*
Manufacturers specified sheet resistance	8–25	15–30	20–40	30–80			ohm/square
Typical measured sheet resistance	8.5	10.75					ohm/square
Measured sheet resistance 10 μm width resistors with side diffusion	8.1	10.3					ohm/square
Measured sheet resistance $V_{gs} = 0V_{ds} = \frac{1}{2}V$					7.7	5.5	kilo-ohm/square
$V_{ds} = 2V$					14.3	9.6	kilo-ohm/square
Measured voltage coefficient	42	69	4.3	9.6			ppm/V
					6.25	10.6	%/V
Measured temperature coefficient	1900	1850	790	680	5200		ppm/°C
Matching	0.4	1.1	0.6	0.8	5		%

A problem with diffused and polysilicon resistors is that the sheet resistances are low in value so that high value resistors occupy a large area. Making the resistors thinner results in a greater spread in value about the desired mean. For diffused resistors, side diffusion comes into play so that the actual width of the resistor not only has to be adjusted to allow for photolithographic errors, but also for the fact that the doping impurities diffuse sideways as well as downwards, thereby increasing the resistor width. The amount of side diffusion depends on the various impurity concentrations but it is typically 0.8 times the depth.

Because of the possible spread in sheet resistance, it is impossible to design circuits with resistors of absolute values. A better approach is to devise circuits that depend on component ratios so that, while the absolute values may change due to process and temperature variations, the ratio remains constant. For precision ratios where end effects caused by having to make a connection to the resistor are important, integral ratios are only possible and must be made using a series of identical resistors. Consider the example given in Figure 2.16(a). A resistor ratio of 3:1 is required. While the centre line length of the longer resistor is three times that of the first, the resistance ratio will not be 3:1. Current crowding occurs in the corner of bends so that the approximate sheet resistance of a corner square falls to about one half. The contact resistance of the connections for the longer resistor is identical to that of the smaller, whereas it should be three times in value to maintain the desired ratio. Thus the correct way to maintain a precise ratio of 3:1 is to make four identical resistors and interconnect them, as shown in Figure 2.16(b).

Where a layout requires resistors to be bent, as illustrated in Figures 2.16(c) and (d), the resistance of the corners must be taken into consideration, particularly if dimensions L_1 and L_2 are not considerably larger than W. Thus for the two cases illustrated,

$$R_{ab} = R_s(L_1/W + L_2/W + 0.56) \quad \textbf{(2.25)}$$

and

$$R_{cd} = R_s/(L_1/W + L_2/W + 0.56) \quad \textbf{(2.26)}$$

While the active resistor formed using a MOS transistor has a much higher sheet resistance (Table 2.3) resulting in smaller area resistors, the transistors are very voltage dependent, particularly the source to substrate voltage. This can be illustrated by a potential divider employing two identical depletion transistors as shown in Figure 2.17(a). SPICE analysis gives the no load output voltage, using typical transistor parameters given in Figure 2.17, as 2.12 volts instead of 2.5 volts—an error of 15.2 percent. The situation can be slightly improved if the transistor gate voltage is increased, thereby reducing the source to substrate voltage effects (Figure 2.17(b)). The output is now 2.44 volts—an error of 2.4 percent.

Earlier it was mentioned that in making contact to a resistor, the actual contact or cut also adds resistance. The amount of resistance varies considerably from foundry to foundry and even from batch to batch. Table 2.4 gives average resistance values of the various minimum size (2λ by 2λ) cuts for two, $\lambda = 2.5\ \mu m$ processes.

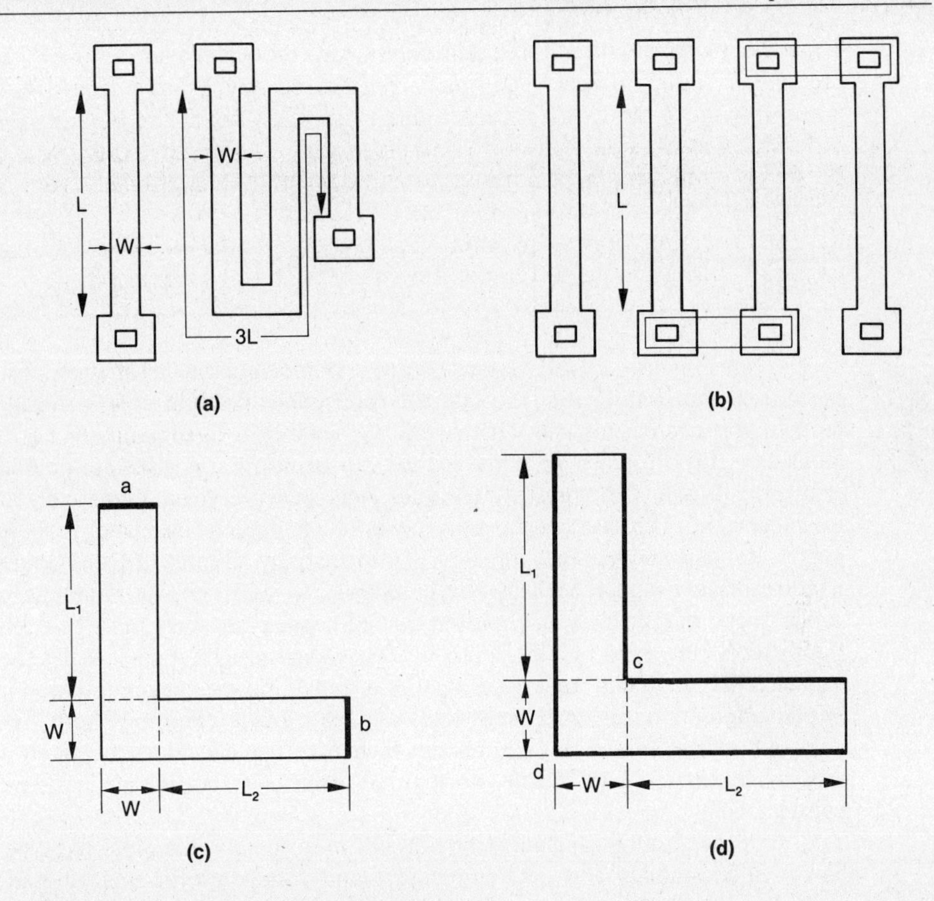

Figure 2.16 Resistor ratios and corners

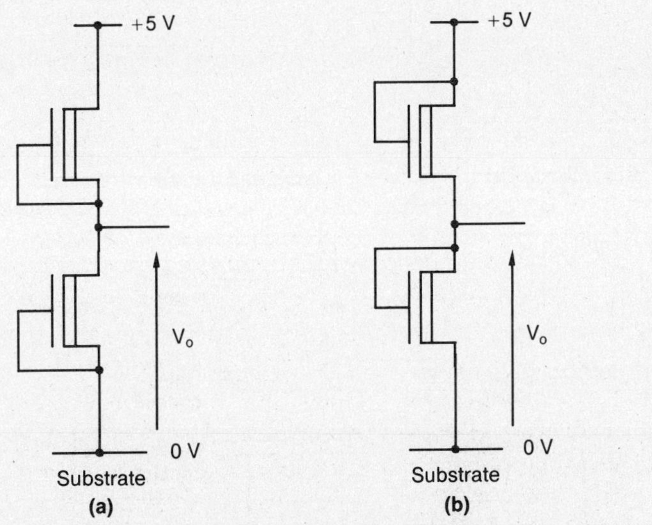

Figure 2.17 Potential divider employing two identical transistors $V_{to} = -3.5$ V, $K = 20\ \mu A/V$, $\gamma = 0.4\ V^{1/2}$ $\phi = 0.7$ V, $\lambda = 0.06$, $W/L = 10/15$

Table 2.4 Average measured resistance for various contacts

Contact or cut type	*Resistance in ohm* Manufacturer 1	Manufacturer 2
Metal to diffusion	4.2	6.4
Metal to polysilicon	7.2	31.4
Polysilicon to diffusion (butting)	25.2	199.4

Inherent in the nMOS process are a number of capacitor types. Important parameters for each of the types are the capacitance per unit area, the voltage and temperature coefficients and whether the capacitance is to substrate or not. For the latter case, if both plates of the capacitor can be above the substrate potential, it is said to be floating. The most useful types together with typical parameters are given in Table 2.5. There are several that are floating, namely the metal to polysilicon, gate to implant and metal to diffusion. Unfortunately all three of them (Figure 2.18) have associated with them a parasitic capacity to substrate. At times the parasitic capacity can be much larger than the desired capacity, so care must be exercised in their use. Of the three types, the last has least to commend it since it has the lowest capacity per unit area and possibly the largest parasitic capacity. Junction (both implant and n^+ diffusion to substrate) and gate oxide capacities both have large voltage coefficients and this can restrict their use. Typical values are given in Table 2.6, while Figure 2.19 graphs normalized values of the voltage coefficients for comparison.

Of the floating capacities, the polysilicon to metal is the preferred type to use, having good stability and low temperature and voltage coefficients. Should only a capacitance to substrate be required, then the metal can be connected to the substrate to approximately double the capacity per unit area, for the polysilicon to metal is in parallel with the polysilicon to substrate. Where voltage effects are of no

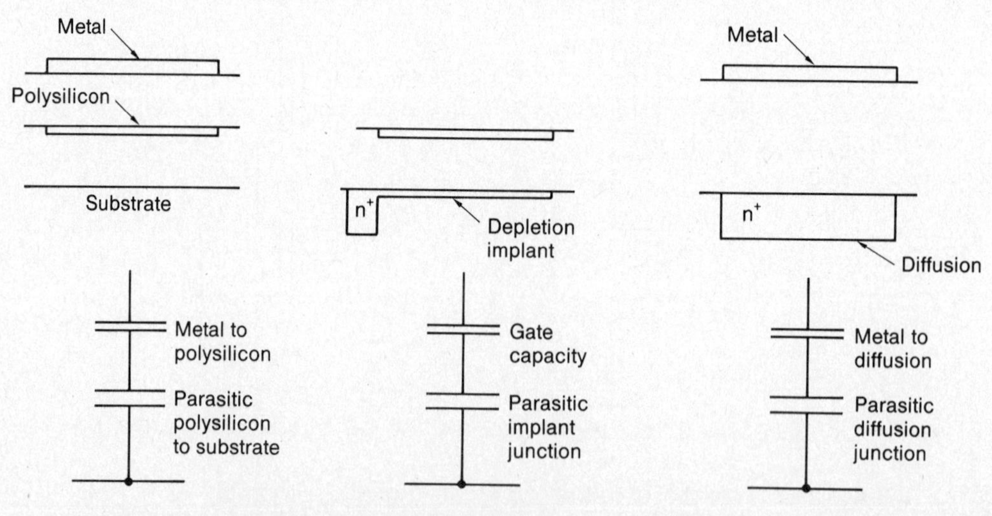

Figure 2.18 Three types of "floating" capacitors

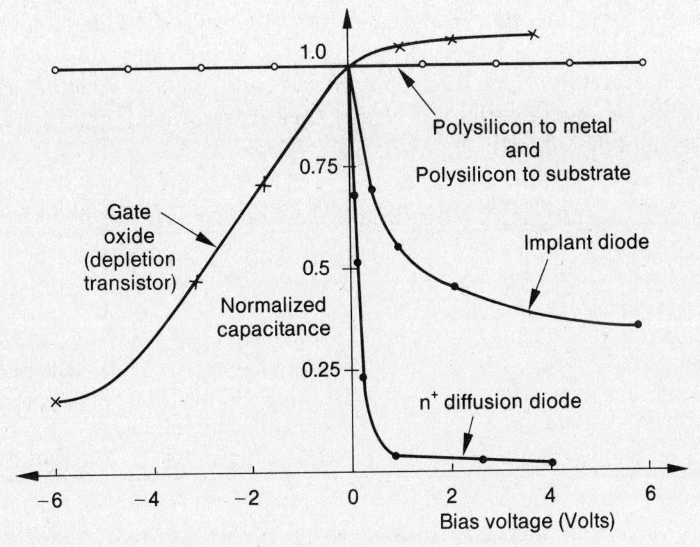

Figure 2.19 Capacitance voltage plots (frequency 10 kHz)

Table 2.5 Typical capacitance values

Parameter	*Manufacture 1* Min.	*Manufacture 1* Max.	*Manufacture 1* Measured	*Manufacture 2* Min.	*Manufacture 2* Max.	*Manufacture 2* Measured	*Units*
Poly to depletion implant	4.1	4.6	4.3	3.6	4.1	3.6	10^{-4} pF/m^2
Temperature coefficient			40			280	ppm/°C
Metal to polysilicon	32	50	46.5			25.5	10^{-6} pF/m^2
Temperature coefficient			50			290	ppm/°C
Polysilicon to substrate	37	47	44.0	43	50	55.5	10^{-6} pF/m^2
Temperature coefficient			120			60	ppm/°C
Junction n diffusion/substrate							
Zero bias area			26			67	10^{-4} pF/m^2
Perimeter			11			4.8	10^{-3} pF/m
+2 V bias area			3.4			4.2	10^{-5} pF/m^2
Perimeter			6.0			5.0	10^{-4} pF/m
Implant to substrate							
Zero bias						26.4	10^{-6} pF/m^2
+2 V bias						11.3	10^{-6} pF/m^2

consequence, the gate oxide capacitor which gives the highest capacitance per unit area is the preferred type. Note Figure 2.19 indicates that for the gate capacity, if the bias voltage is greater than 2 volts, and for the junction capacity, if the bias voltage is greater than 1 volt, then capacitance changes with voltages are considerably reduced.

Table 2.6 Typical voltage coefficients for the various capacitor types

Capacitor type	*Typical voltage coefficient*	*Comment*
Polysilicon to depletion implant	2.10^5 ppm/volt	For bias voltages >2 V
Metal to polysilicon	980 ppm/volt	Independent of bias voltage
Polysilicon to substrate	−2700 ppm/volt	Independent of bias voltages
n^+ diffusion to substrate	-1.8×10^6 ppm/volt	For bias voltage >1 V
Implant to substrate	-4.8×10^4 ppm/volt	For bias voltages >1 V

Various methods of achieving accurate capacitor ratios have been reported (McCreary and Gray, 1975, pp. 371–9; Newcomb, 1981, pp. 40–2), for, as with resistors, it is preferable not to depend on an absolute value of a component, but a component ratio. The capacitors should be designed to allow for the oxide layers not being of uniform thickness. An example is given in Figure 2.20(a) for a capacitor ratio of 2:1. While the idea has merit, it is often difficult to implement because the added capacity formed by the interconnection links can be larger than the variations caused by the nonuniform thickness of the oxide. This then leads to the second and far more important point that all interconnection tracks must be considered as part of the capacitor. Thus, in making precision capacitor ratios, the actual interconnection track must be considered as part of a unit capacitor cell. See Figure 2.20(b).

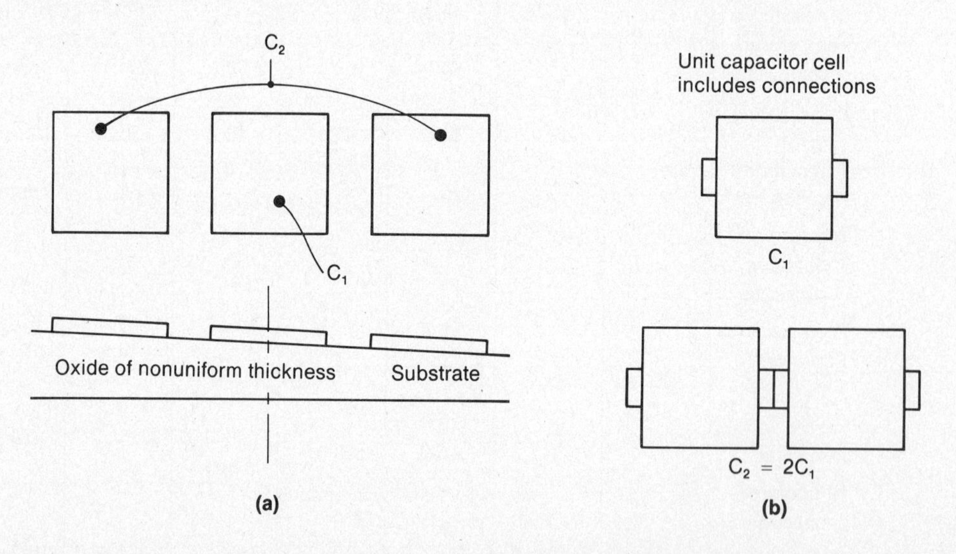

Figure 2.20 Methods to assist in producing precision capacitor ratios

2.7 Component noise

An important part of some analog circuit designs is the signal to noise ratio. Consider Figure 2.21. An ideal signal (v_i) is applied to a nMOS amplifier to be amplified. The output from the amplifier consists, in addition to the amplified signal (Av_i), of a noise component (v_n) that has been generated by the amplifier itself. We can refer this noise component back to the amplifier input to give an equivalent input noise signal of v_n/A. It is this value that sets the lower limit to the magnitude of the desired signal v_i that can be amplified.

Just what components in an nMOS amplifier generate noise? Capacitors tend to produce little or no noise. Small fluctuations in the dielectric of electrostatic doublets do give rise to minute capacity variations, but these are insignificant compared to the noise generated by other components. For example, all resistors generate Johnson or thermal noise due to the random thermal motion of the charge carriers. It is a white noise in that it has a constant power density independent of frequency and can be expressed either as a noise current or voltage form, as shown in Figure 2.22. For a 1 k-ohm resistor, at 27°C the noise voltage/$\sqrt{\text{Hz}}$ is $4.06.10^{-9}$, which for most nMOS circuits is small and can be ignored.

The remaining component is the transistor. It has at least two major noise sources (Figure 2.23). The first is the thermal channel noise because the channel resistance of a transistor is finite. For a transistor in the saturation region, we have (Van der Ziel, 1970):

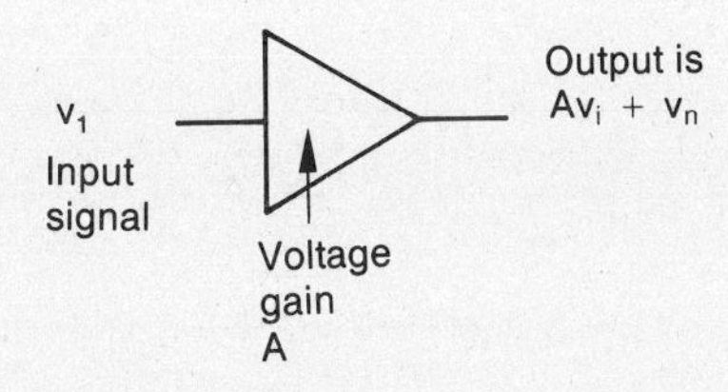

Figure 2.21 Amplifier with noise generation

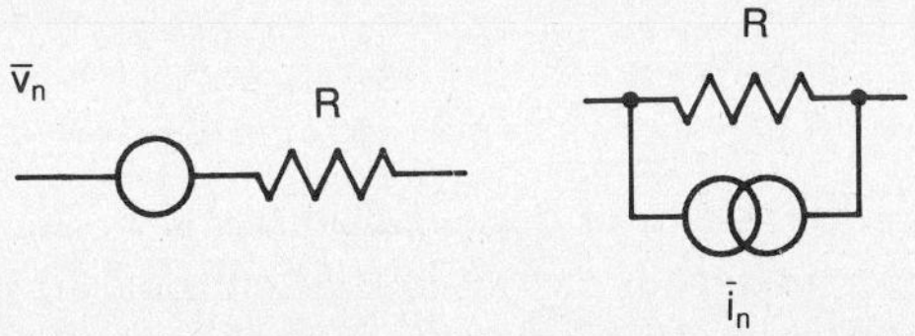

Figure 2.22 Thermal noise of a resistor R

Notes:

$\overline{v_n^2} = 4kTR\Delta f$ $\qquad \overline{i_n^2} = 4kT\Delta f/R$

k = Boltzman's constant

T = temperature °K

R = resistance in ohm

Δf = the bandwidth

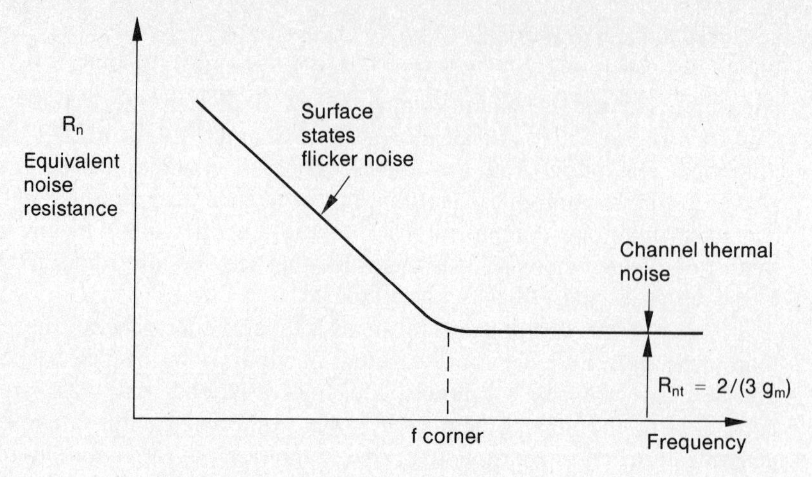

Figure 2.23 Input noise resistance for a nMOS transistor as a function of frequency

$$(\bar{i}^2_{dnt}) = 4kT\frac{2g_m}{3}\Delta f \tag{2.27}$$

where $\bar{i}_{dnt}$ is the drain thermal noise current.

For convenience it is usual to express the noise as a gate voltage (v_{gnt}) in terms of an equivalent noise resistance R_{nt}

where

$$\bar{i}_{nt} = g_m v_{gnt}$$

Thus,

$$\overline{v}^2_{gnt} = 4kTR_{nt}\Delta f \tag{2.28}$$

where

$$R_{nt} = \frac{2}{3g_m} \tag{2.29}$$

The second source of noise, called *flicker noise*, is generated by surface states. While there are numerous expressions attempting to accurately describe this noise, the one selected here is by Klassen (Van der Wiele et. al., 1977) namely,

$$\bar{i}^2_{dnf} = \frac{K_1 g_m I_d}{C_g f}\Delta f \tag{2.30}$$

where K_1 is a proportionality constant and C_g is the gate capacity and equals

$$\frac{\epsilon WL}{D} = C_{ox}WL$$

Because this noise is inversely proportional to frequency, it is certainly the dominant noise at low frequencies. However, even with modern nMOS technology, the corner point frequency (Figure 2.23), where flicker noise becomes equal to the thermal noise, can be as high as 1 MHz. Thus, for normal analog operation, flicker noise is most important.

Equation 2.30 shows how to minimize flicker noise. For the saturation region of operation we have, using equation 2.13,

$$g_m = \left.\frac{dI_d}{dV_{gs}}\right|_{V_{ds} = \text{constant}}$$

$$= \mu C_{ox}\frac{W}{L}(V_{gs} - V_{to})$$

or

$$g_m = \left(2\mu C_{ox}\frac{W}{L}I_d\right)^{\frac{1}{2}} \tag{2.31}$$

Substituting into equation 2.30,

$$\bar{i}_{dnf}^{\,2} = \sqrt{\frac{2\mu K_1^4 I_d^3 \Delta f^2}{WL\, C_{ox} f^2 L^2}} \tag{2.32}$$

From equation 2.32 we see that to minimize flicker noise, the device should be run at a low drain current and that the area (WL product) should be made as large as possible. The remaining term that is under the designer's control, device length (L^2 term), must be treated with care as increasing L reduces the stage gain. Thus the front stages of a low noise amplifier should be of large area and operate at low current levels.

2.8 nMOS lambda based layout rules

As mentioned in Chapter 1 part of the standard interface to fabrication are the integer lambda based layout rules. These are given in Figure 2.24 and Plate 1.

2.9 Exercises

1 An enhancement transistor is connected as shown in Figure 2.25. If the threshold voltage can vary from 0.6 volts to 1.0 volts and K from 20 to 25 μA/V, calculate the spread in drain current assuming the simple models of equations 2.12 and 2.13. State any assumptions made.

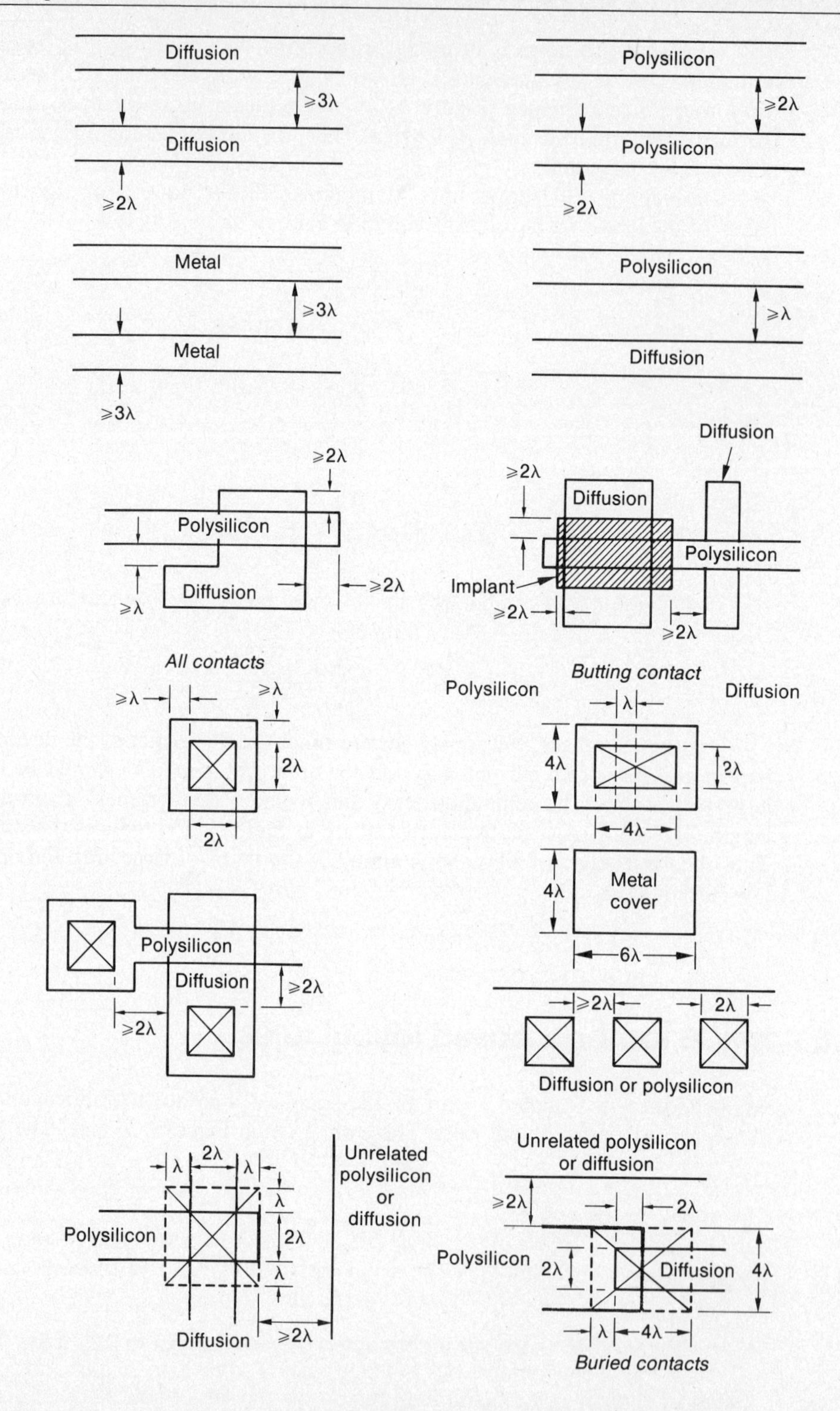

Figure 2.24 Lambda based layout rules

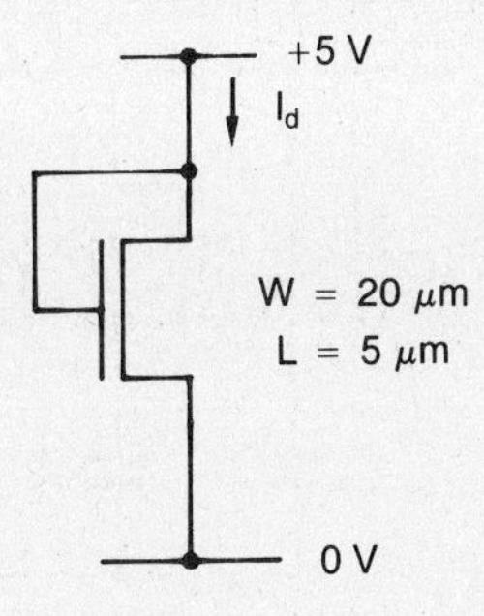

Figure 2.25

2 Using the Lambda based rules given in Figure 2.24, complete the layout of a 1 kilo-ohm resistor, to occupy minimum area using:
(a) N^+ diffusion of 10 ohm/square;
(b) polysilicon of 20 ohm/square;
(c) a depletion transistor of 8 kilo-ohm/square.

3 Refer to Figure 2.26. Using the capacitance figures measured for Manufacturer 1 (Table 2.5), calculate the input gate and drain to substrate capacities.

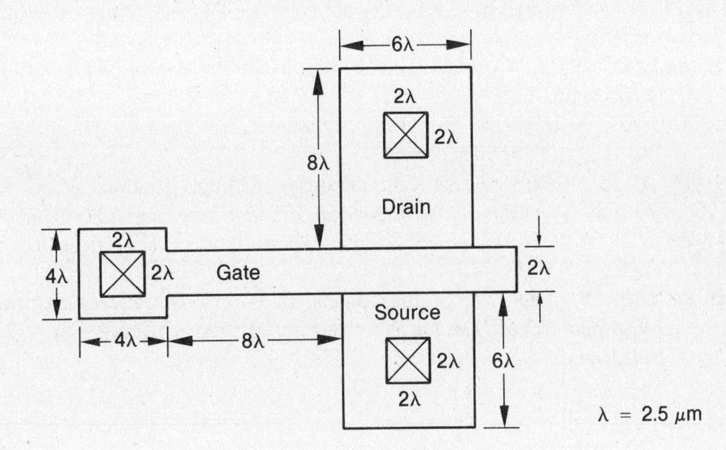

Figure 2.26

4 Accurate ratioed binary weighted capacitors of C, 2C, 4C and 8C are required and are to be made using the metal to polysilicon capacitor. Each capacitor is multiplexed (as shown in Figure 2.27). Devise several ways of laying out this system using 15 unit circuits. On which side of the capacitors would you suggest placing the parasitic capacitor?

5 For the inverter amplifier given in Figure 2.10, verify the transfer characteristic given in Figure 2.11. What is the output DC voltage under self bias conditions, that is, when the output voltage equals the input?

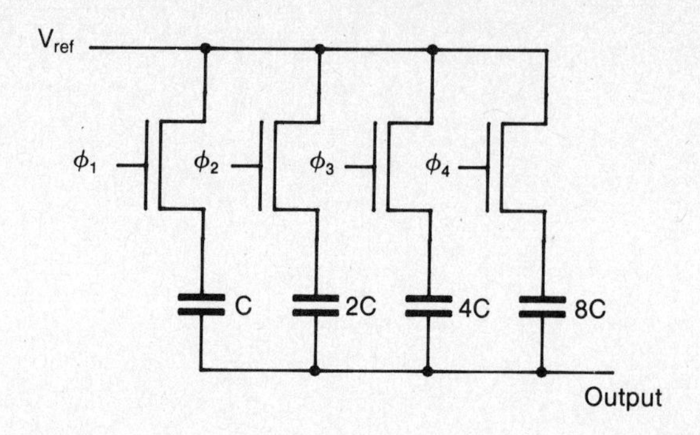

Figure 2.27

2.10 References

Ambrozy, A. (1982), *Electronic Noise*, McGraw-Hill International, New York.

McCreary, J. L. and Gray, P. R. (1975), "All MOS Charge Redistribution Analogue to Digital Conversion Technique", *IEEE Journal of Solid State Circuits*, Part 1, Vol. SC–10, pp. 371–9.

Mead, C. and Conway, L. (1980), *Introduction to VLSI Systems*, Addison Wesley, Reading, Mass.

Newcomb, R. (1981), "Considerations in MOS Ratioed Capacitor Layout, *VLSI Design*, Third Quarter, pp. 40–2.

Pucknell, D., Eshraghian, K. (1985), *Basic VLSI Design: Principles and Applications*, Prentice–Hall, Sydney.

Tsividis, Y. P. (1978), "Design Considerations in Single Channel MOS Analogue Integrated Circuits—A Tutorial", *IEEE Journal of Solid State Circuits*, Vol. SC–13, pp. 383–91.

Van der Ziel, A. (1970), "Noise Sources, Characteristics and Measurement", Prentice–Hall, Englewood Cliffs, N.J.

Van de Wiele, F., Engl, W. L. and Jespers, P. G. (1977), *Process and Device Modelling for Integrated Circuit Design*, Nato Advanced Study Institute Series, Noordhoff, Leyden, Netherlands.

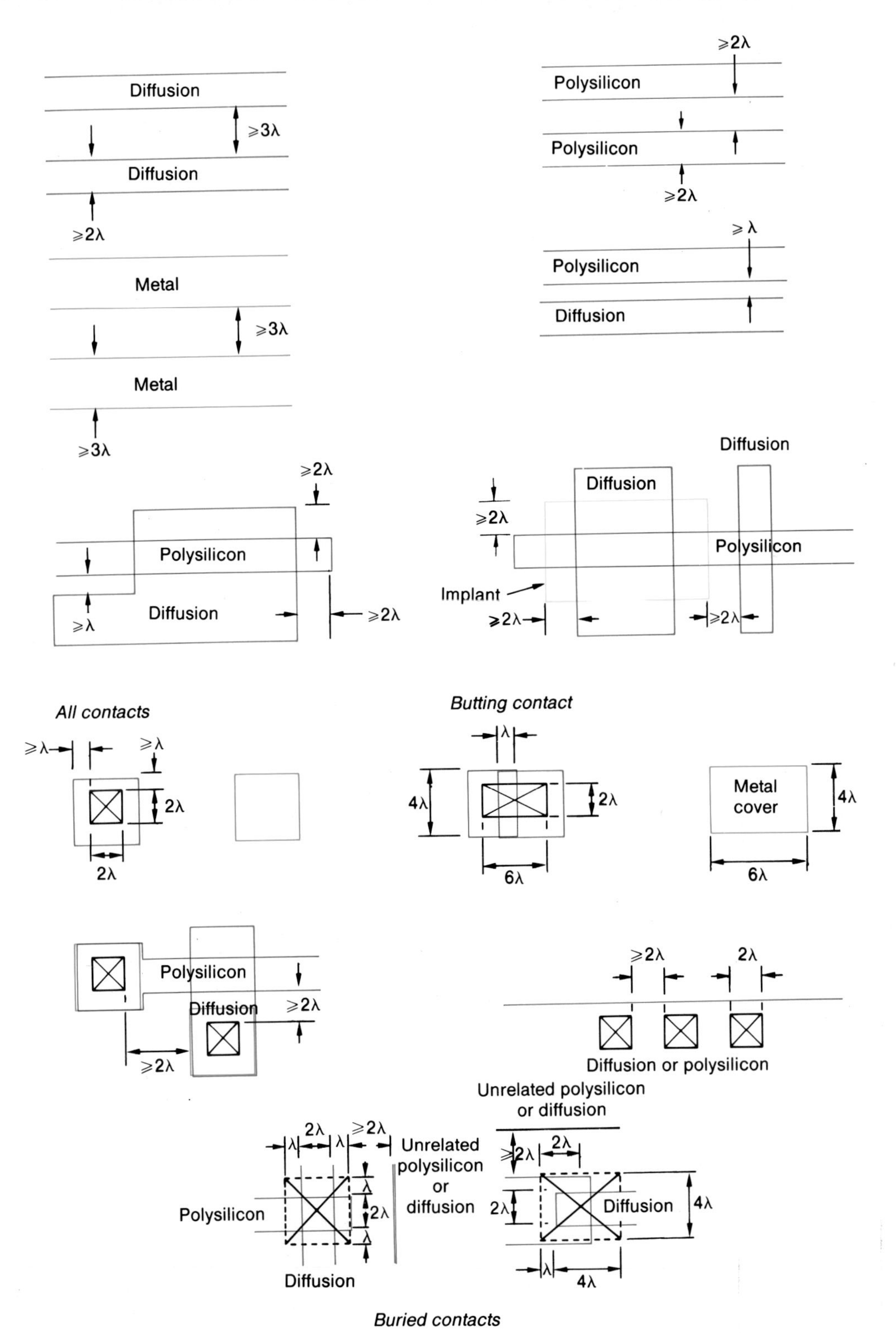

Plate 1 nMOS lambda based layout rules

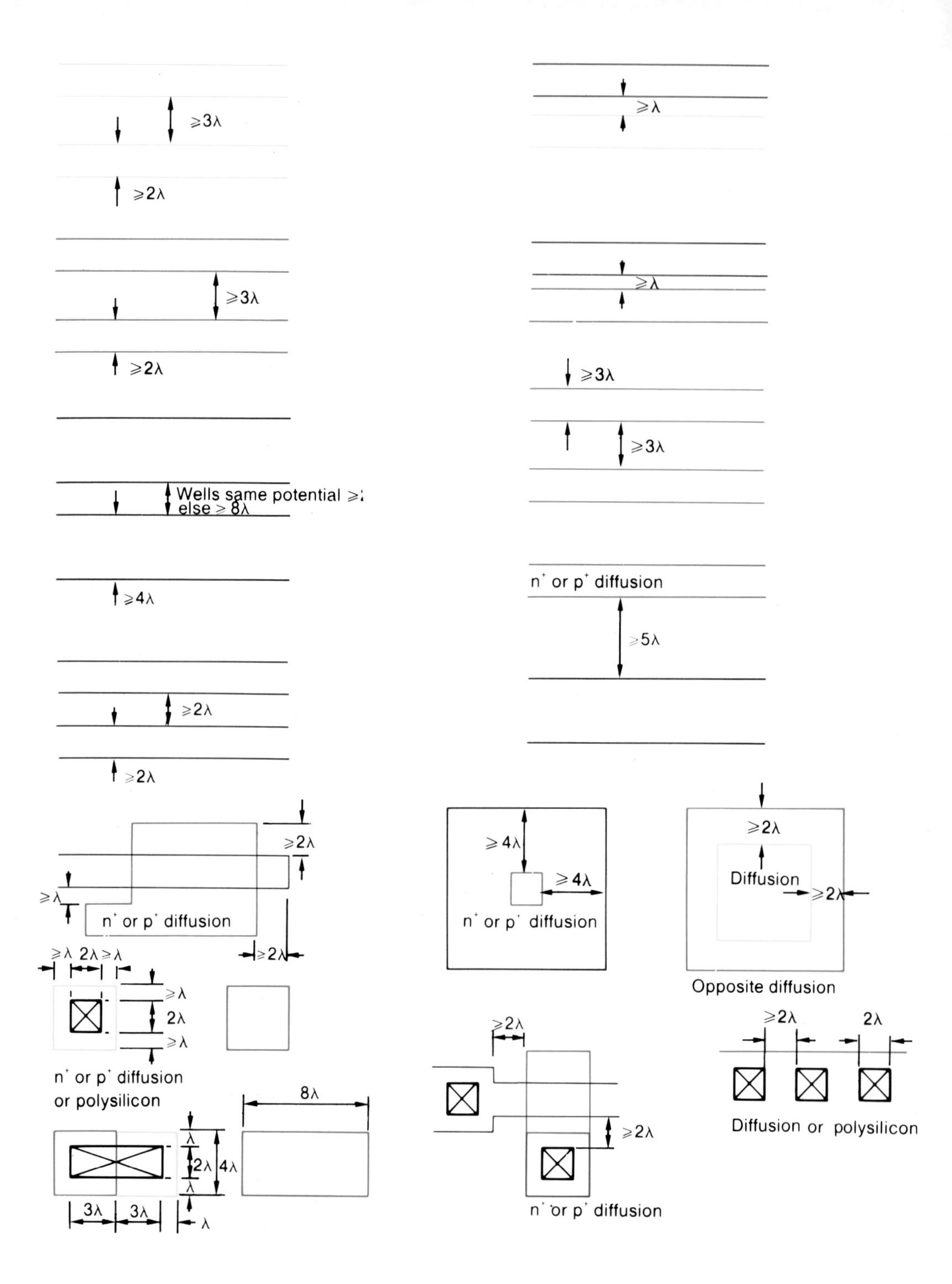

Plate 2 Portable CMOS lambda based layout rules

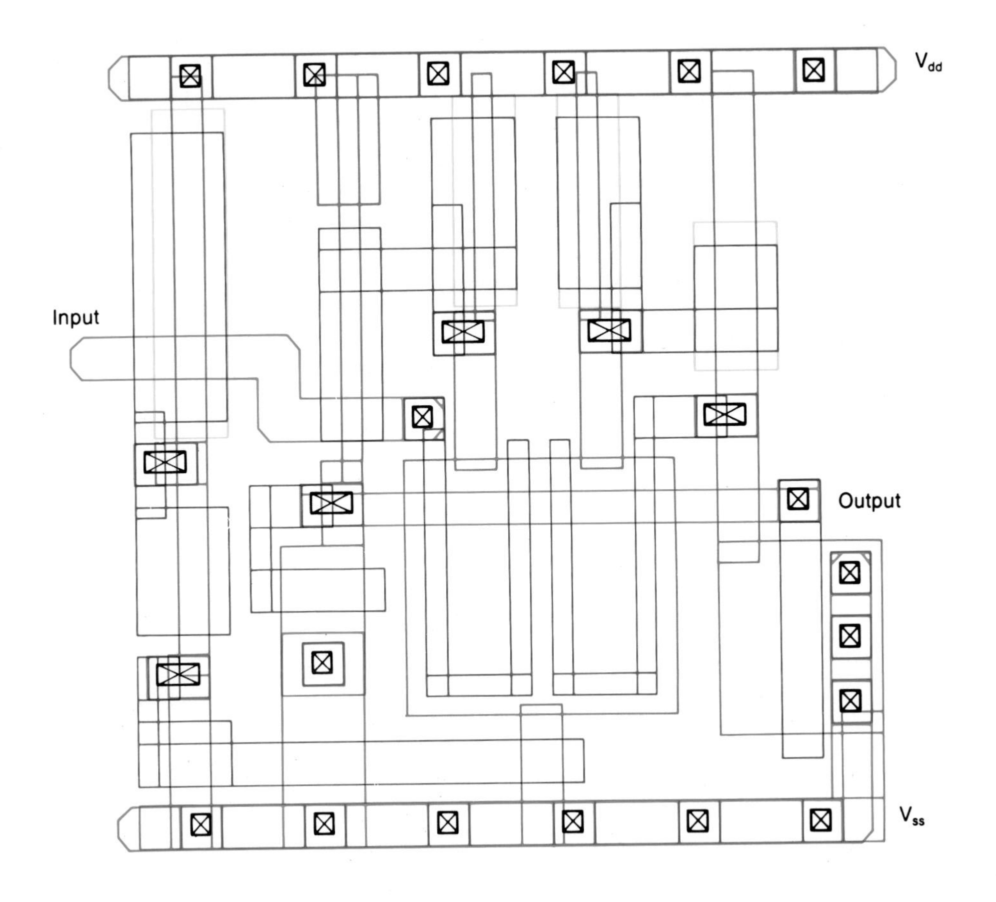

Plate 3 A unity gain nMOS buffer amplifier (BUFAMP)

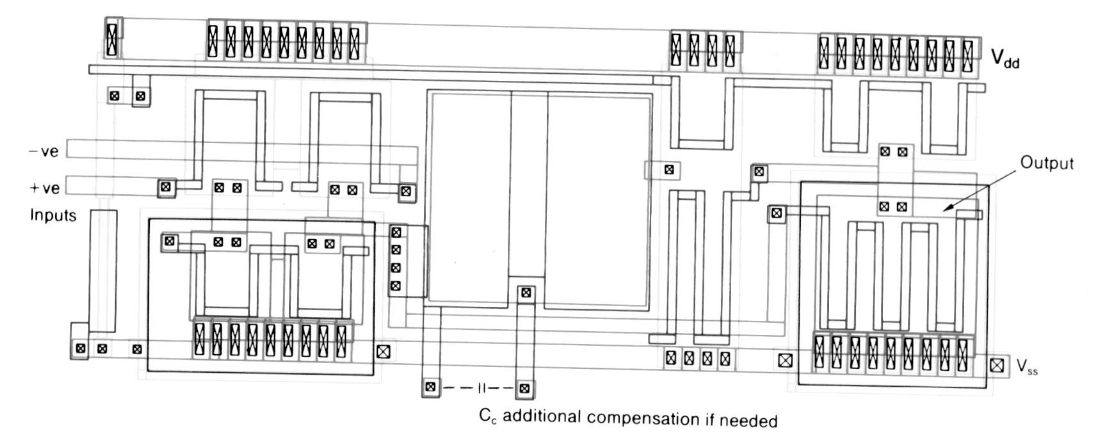

(a)

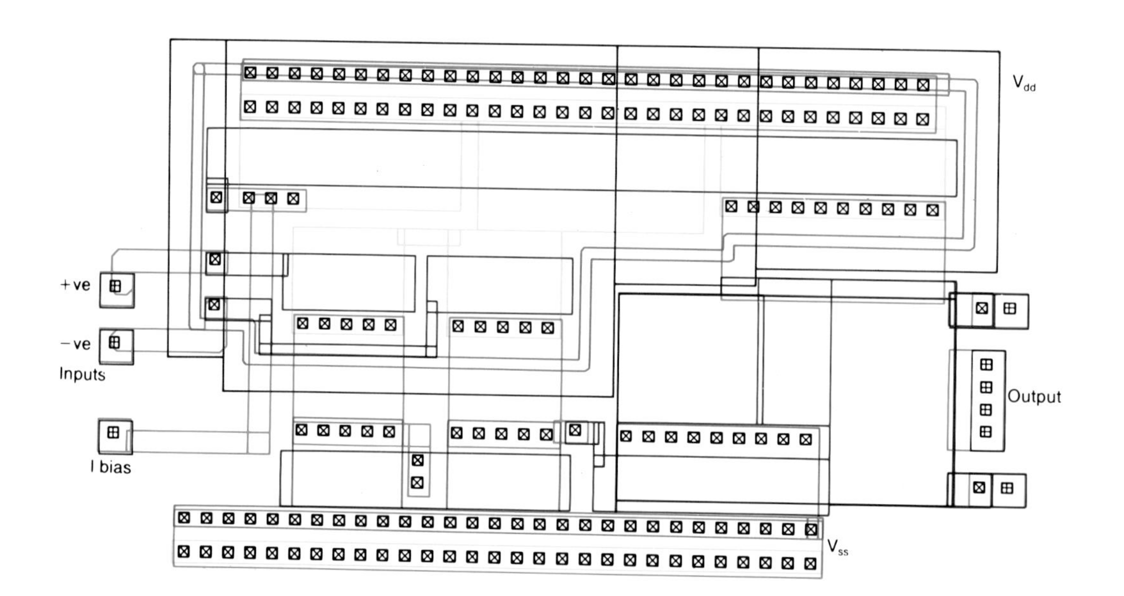

(b)

Plate 4 CMOS operational amplifiers for **(a)** p-well and **(b)** n-well processes

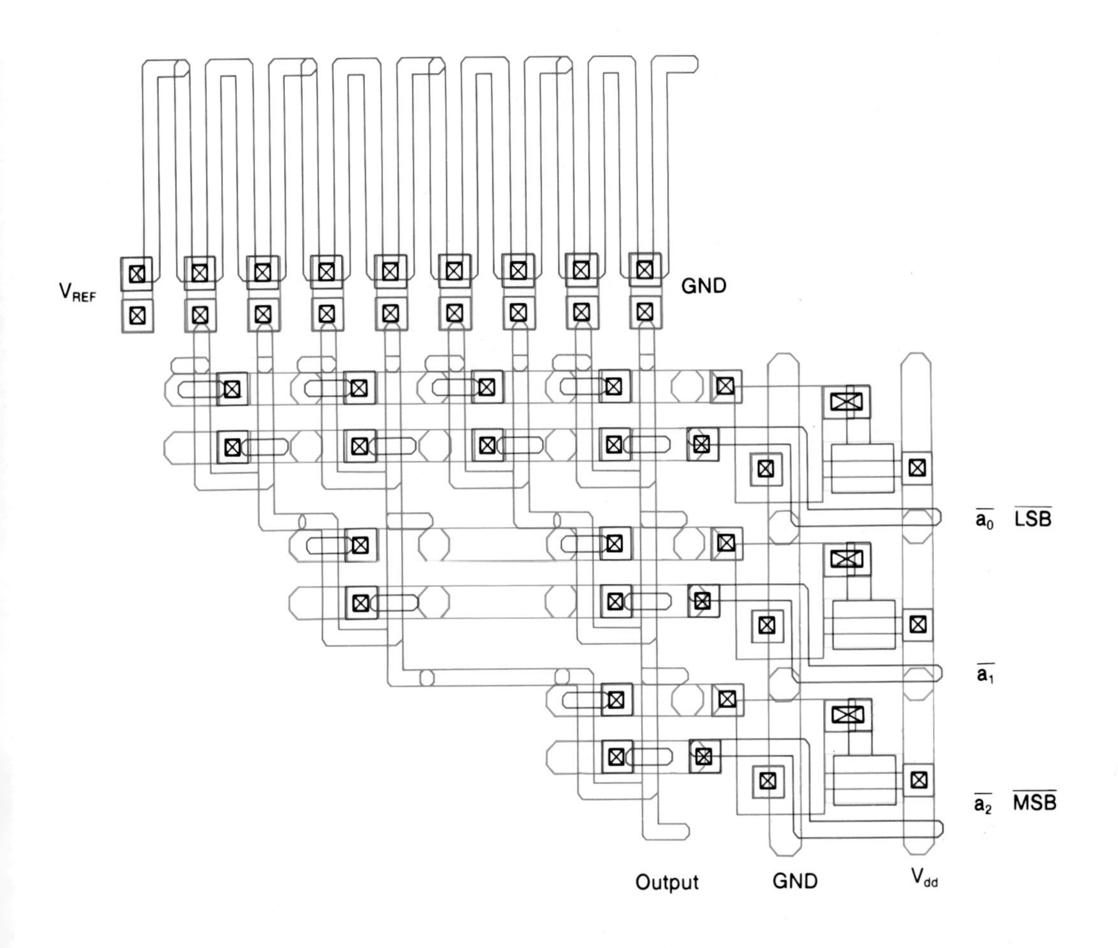

Plate 5 Layout of a three-bit resistor string converter

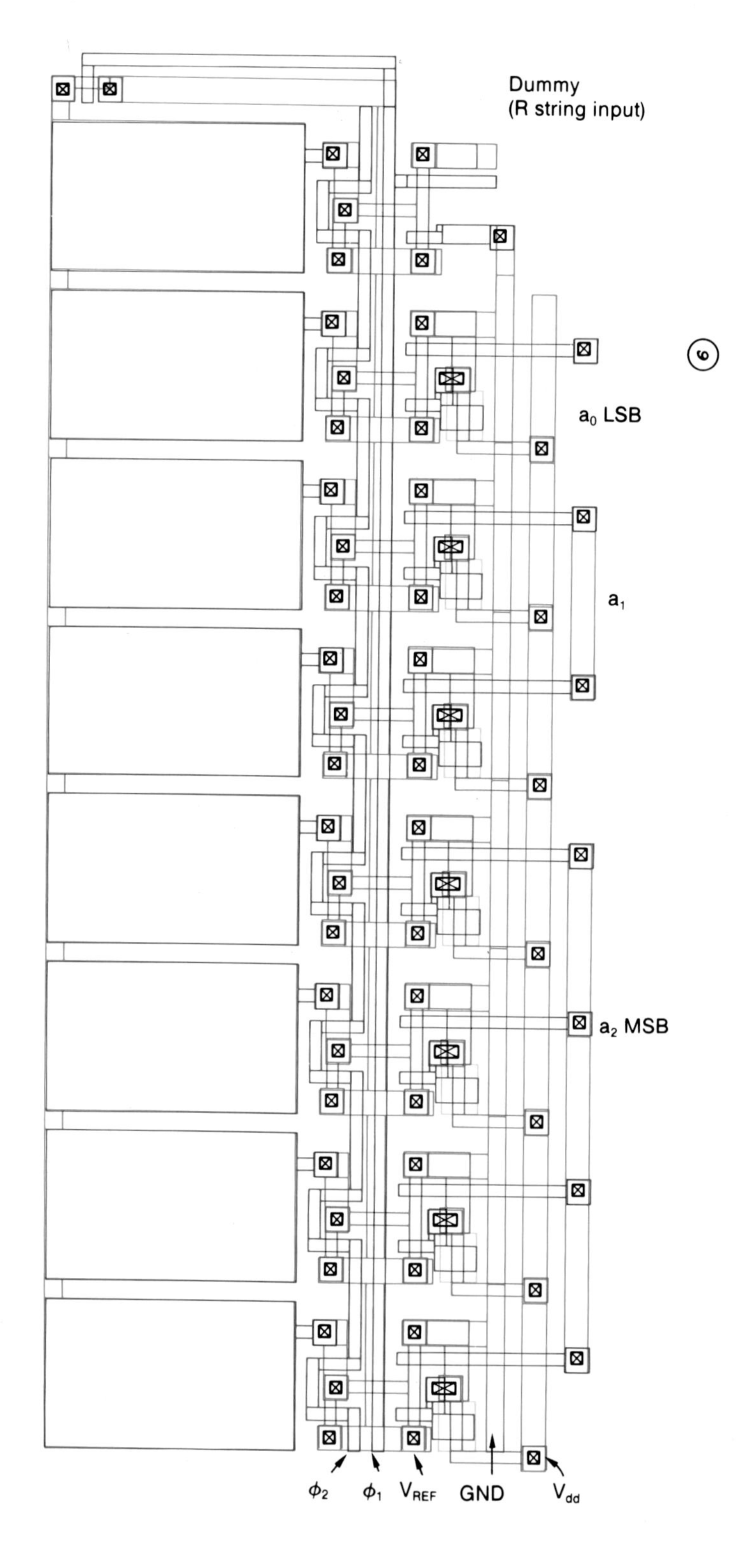

Plate 6 Charge redistribution D/A converter

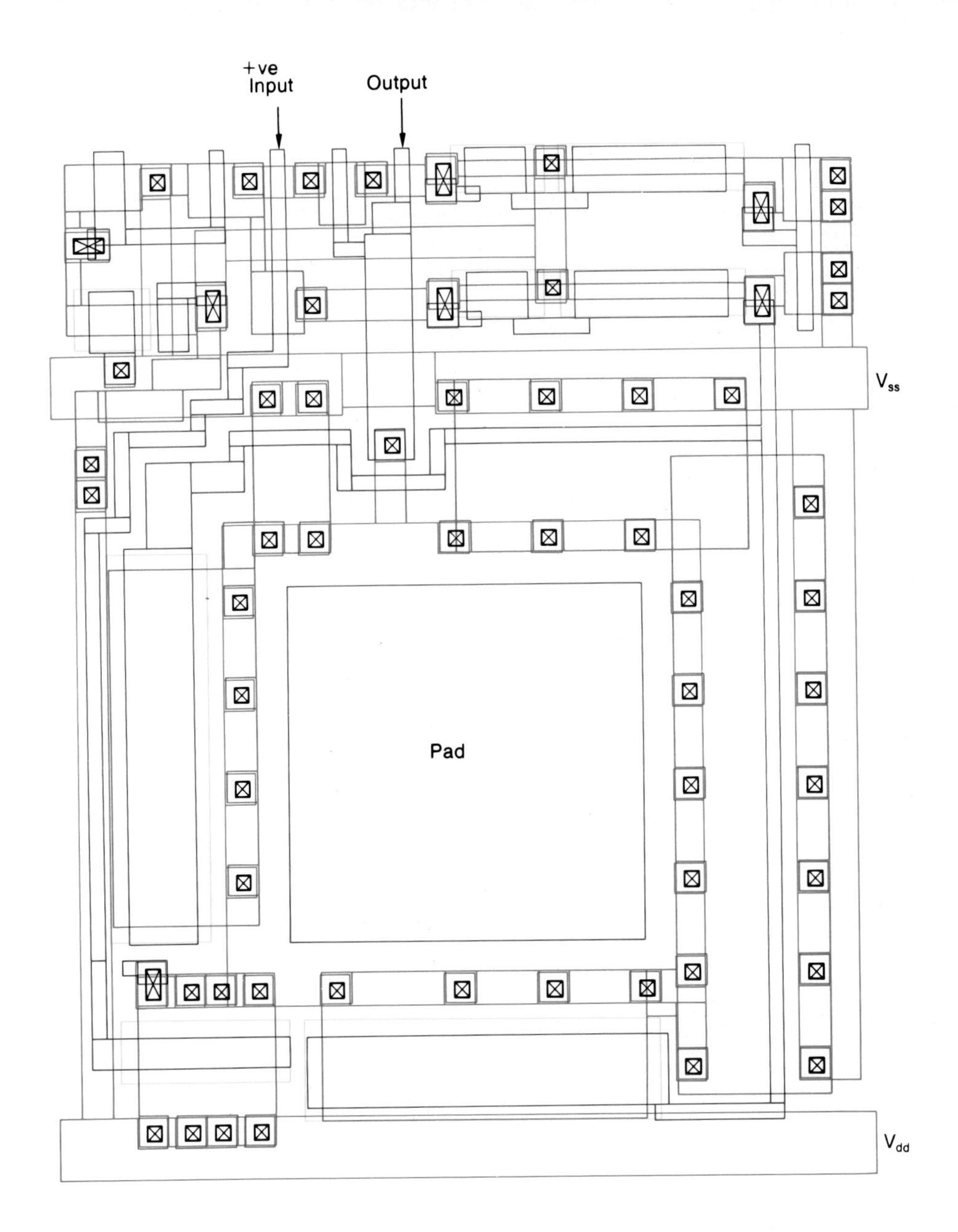

Plate 7 nMOS analog output pad

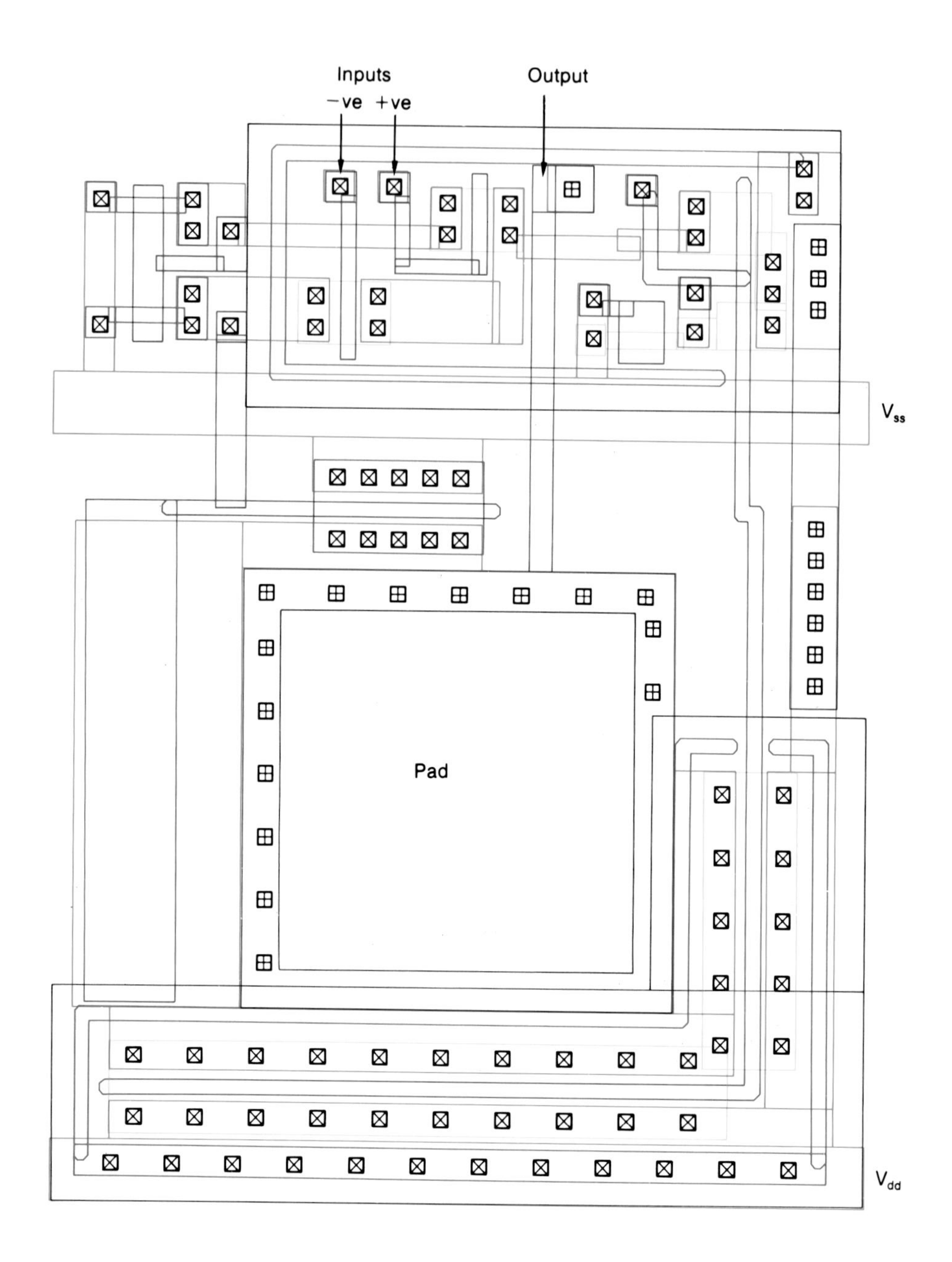

Plate 8 CMOS analog output pad

3 Components of the CMOS process

3.1 Introduction

Unlike nMOS, where most manufacturers have a very similar process, there are many variations for CMOS. These include p-well, n-well, twin well or tub (both p-and n-well), silicon on insulator (such as sapphire), oxide isolation, metal or polysilicon gate and double polysilicon or metal (Weste and Eshraghian, 1985). Some of these variations are illustrated in Figure 3.1. As a consequence, it is more difficult to present a simple unified approach.

The simplest process was the original bulk p-well metal gate process and, despite its age, it is still used for many products. A silicon gate version is possibly the most common process at present, although the n-well and twin tub processes are gaining in popularity. These are all bulk processes and are not unlike the nMOS technology previously discussed except that the two MOS transistors available are complementary enhancement types, that is, n- and p-channel. Unfortunately, when a single well is used, the devices made in the high resistance wells tend to have a much larger back voltage effect, gamma (γ). Further, because of the difference in mobility of electrons and holes, the conduction factor (K) for the p-channel devices is about a half to a third of that of the n-channel MOS.

An advantage of the n-well process is that with minor changes nMOS chips can be produced on the same line. The twin tub process allows the trimming of the characteristics of both transistor types and, with simple changes, can revert to either a bulk p- or n-well process. Oxide isolation and silicon on insulator (SOI) processes reduce capacities and therefore give improved speed. Double metal or polysilicon

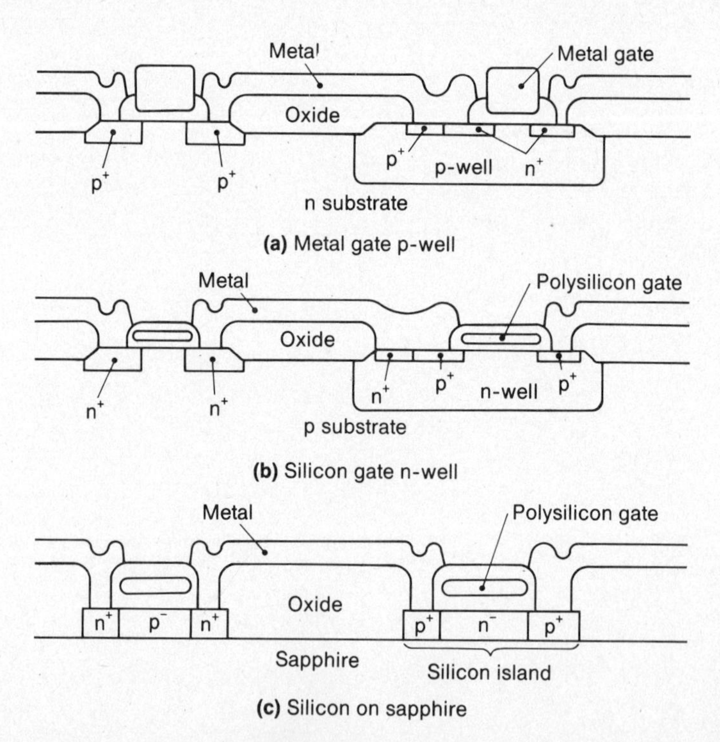

Figure 3.1 Sample CMOS processes

layers provide greater design flexibility including the design of switched capacitor circuits. Thus, all the various process variations are to provide either a more flexible or a superior end product. However, in the main, the processing steps to produce CMOS circuits are similar to those already discussed for nMOS and therefore only a simplified description for the p-well process will be given. This has the advantage of allowing a clearer view of just what masks are required for a process and how they can be derived from a unified design layout.

3.2 The silicon gate p-well process

In this process an n-type wafer of 2–5 ohm cm resistivity is employed as the starting material. For single layers of polysilicon and metal, eight masks (given in Table 3.1) are required for the process. Names and colors suggested by the Jet Propulsion Laboratory, California Institute of Technology, are also given (Grisworld, 1982). Many other colors and standards have been proposed (Weste and Eshraghian, 1985; Lipman, 1981; Grisworld, 1982; Natural Sciences and Engineering Research Council of Canada, 1985; Joint Microelectronic Research Centre, 1985). Note that the table only lists seven masks as the fifth mask used in the process can be derived from information contained in the other masks. This will be discussed later.

Before considering the manufacture of a simple CMOS inverter (Figure 3.2), some comments on substrate and well biasing need to be made. Since the p-well

Table 3.1 Mask layers and representation for the p-well process

Mask layer	*Plot color*	*Plot/mask code name*	*Comment*
p-well	Brown	CW	
thin oxide	Green	CD	Defines where p- and n-transistors are.
Polysilicon	Red	CP	
p^+	Yellow	CS	Differentiates where p-transistors are to be.
Contact	Black	CC	
Metal	Blue	CM	
Overglaze	—	CG	Often made brown as for nMOS.

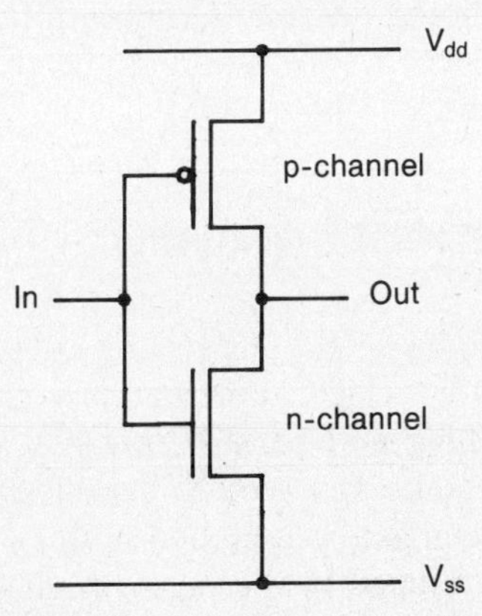

Figure 3.2 Simple CMOS inverter

must be isolated from the n-type wafer, the well must be reverse biased. Thus it is usual to connect the p-well to the most negative voltage (V_{ss} which is usually 0 volts) and the substrate to the most positive voltage (V_{dd}). These contacts are also included in this inverter example and are shown in Figure 3.3.

A field oxide of 0.5–1 μm thickness is grown and used in the photolithographic process. With reference to Figure 3.3, the first mask (brown) is used to define the p-well areas, a high resistance diffusion typically 5 kilo-ohm per square and 4 μm deep. Next, the second mask, thin oxide (green), defines the areas where transistors are to be made—transistors of both n- and p-channels. The thin oxide is typically .08 μm thick. In the nMOS process, all transistors were n-type. Consequently, this mask is equivalent to the diffusion mask of the nMOS process. The color is the same—green.

Following the defining of transistor areas, the polysilicon layer is deposited, using the third mask (red) in the series, to pattern the areas. When the p- and n-implants occur (Masks 4 (yellow) and 5), the masking action of the polysilicon results in self-aligned gates. As stated previously, the fifth mask for the n-implant is generated from information on other masks. In the simple case under consideration here, it is the complement of the p-implant Mask 4. In other processes more complex logical operations may be needed. To make transistors with the p-well process, an nMOS transistor must be surrounded by a p-well, whereas a pMOS transistor must be surrounded by a p-plus region. Thus the p-implanted areas of silicon are derived by the logical AND operation of the thin oxide (Mask 2) and p^+ (Mask 4), whereas the n-implanted areas of silicon are derived by the logical AND operation of the thin oxide (Mask 2) and the complement of the p^+ regions (Mask 5).

The final three masks define the position of contact holes or cuts, the metal interconnect pattern and the overglaze layer. The metal mask is often of opposite polarity to the other masks as the mask defines where metal is to be left rather than where the field oxide is to be removed. Thus, with eight masks, the inverter circuit, together with substrate and well connections, has been fabricated. We will now examine in more detail the characteristics of the devices that can be produced. Because both n- and p-diffusions are used, a number of devices other than the two MOS transistors are available, and since they are advantageous for analog work, they will be discussed in this chapter.

It should be noted that the n- and p-diffusions make it possible to have top connections to the silicon substrate and therefore one does not rely on a back connection as was the case with nMOS.

3.3 The enhancement transistor

As previously mentioned, the CMOS process offers both n- and p-channel enhancement transistors which, while complementary, are far from being matched. The degree of matching and parameter values depend on the particular CMOS process. Table 3.2 compares typical parameter values for three different bulk processes, namely metal gate p-well, and both p- and silicon gate n-well. The MOS transistor models developed in the previous chapter for the nMOS process apply equally here. One must remember to reverse both the voltage polarities and direction

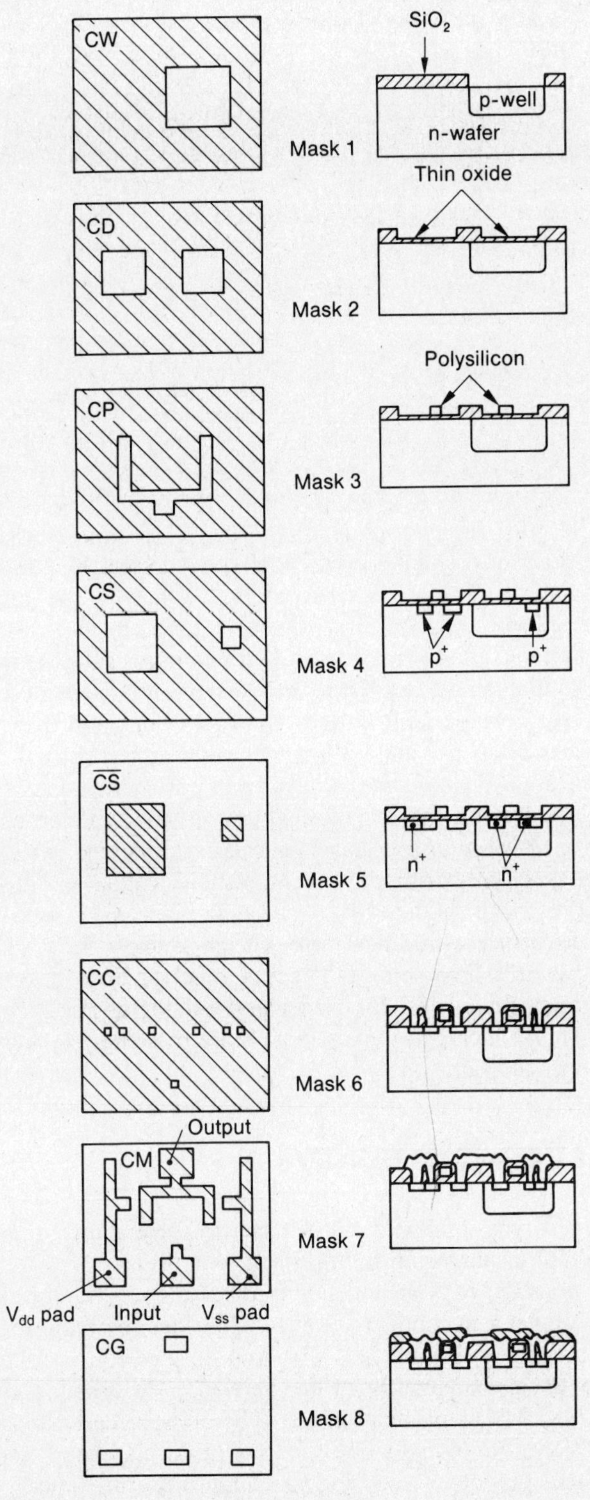

Figure 3.3 Steps in the p-well CMOS process

Table 3.2 Typical SPICE parameters for three different CMOS processes

Transistor type	*Parameter*	*Metal gate p-well*			*Si gate p-well*			*Si gate n-well*			
		Min.	Typ.	Max.	Min.	Typ.	Max.	Min.	Typ.	Max.	*Units*
Enhancement n-channel	V_{to}	0.6	1.0	1.3	0.5	0.7	1.0	0.5	0.7	0.9	V
	K		15			30		20	24	27	$\mu A/V^2$
	λ	0.01	0.015	0.03	0.025		0.04		0.015		V^{-1}
	γ		1.35		1.0		1.2	0.1		0.2	$V^{\frac{1}{2}}$
	x_j		2.5			1.0			0.5		μm
	phi		0.64			0.65			0.65		V
Enhancement p-channel	V_{to}	−1.3	−1.0	−0.6	−1.0	−0.7	−0.5	−0.9	−0.7	−0.5	V
	K		7			11		7	8.5	10	$\mu A/V^2$
	λ	0.01	0.02	0.05	0.005		0.01		0.05		V^{-1}
	γ		0.5		0.4		0.6	0.3		0.5	$V^{\frac{1}{2}}$
	x_j		2.5			1.0			0.5		μm
	phi		0.64			0.65			0.65		V

of current flow for p-channel transistors. To readily distinguish between symbols for p- and n-channel devices, either a circle is added to the gate for the p-device or arrows are added to the sources of both transistors. Both are illustrated in Figure 3.4. In this book, the former method will be used. As explained in Chapter 2, it is not usual to include the back gate on the symbol but any design analysis must take the substrate or well potentials into consideration. Figure 3.4 also shows the static characteristics for n- and p-channel MOS transistors. Substrate and well connections are shown in this figure but will not be included in future.

Before leaving the enhancement transistor types, it should be pointed out that some CMOS processes have natural transistors or possibly several enhancement transistor types with different threshold voltages. Thus, in addition to the standard types (with characteristics similar to those given in Table 3.2), there may be an n-channel device with a near zero gate threshold voltage and a p-channel with a more negative threshold voltage. The n-channel device is called the *natural transistor* of the process and is ideal for pass transistor logic. The p-channel device may be more suitable for memory applications. Since these devices are not standard, we will not refer to them again.

3.4 Bipolar transistor theory

In addition to the enhancement MOS transistors, one or two types of bipolar transistors are available, depending on the process. As the name implies, this transistor consists of two junctions formed by a sandwich of three alternate oppositely doped semiconductor layers. Consider the npn bipolar transistor shown diagrammatically in Figure 3.5. The middle p-region is called the base and, for reasons to be given later, must be narrow in width if high gains and a good frequency response are to be obtained. The outer n-regions may not be symmetrical as shown here. In discrete devices, the physically larger n-region is called the *collector* and the other the *emitter*. Under normal linear operation, the base collection junction diode is reverse biased so that the collector diode has to

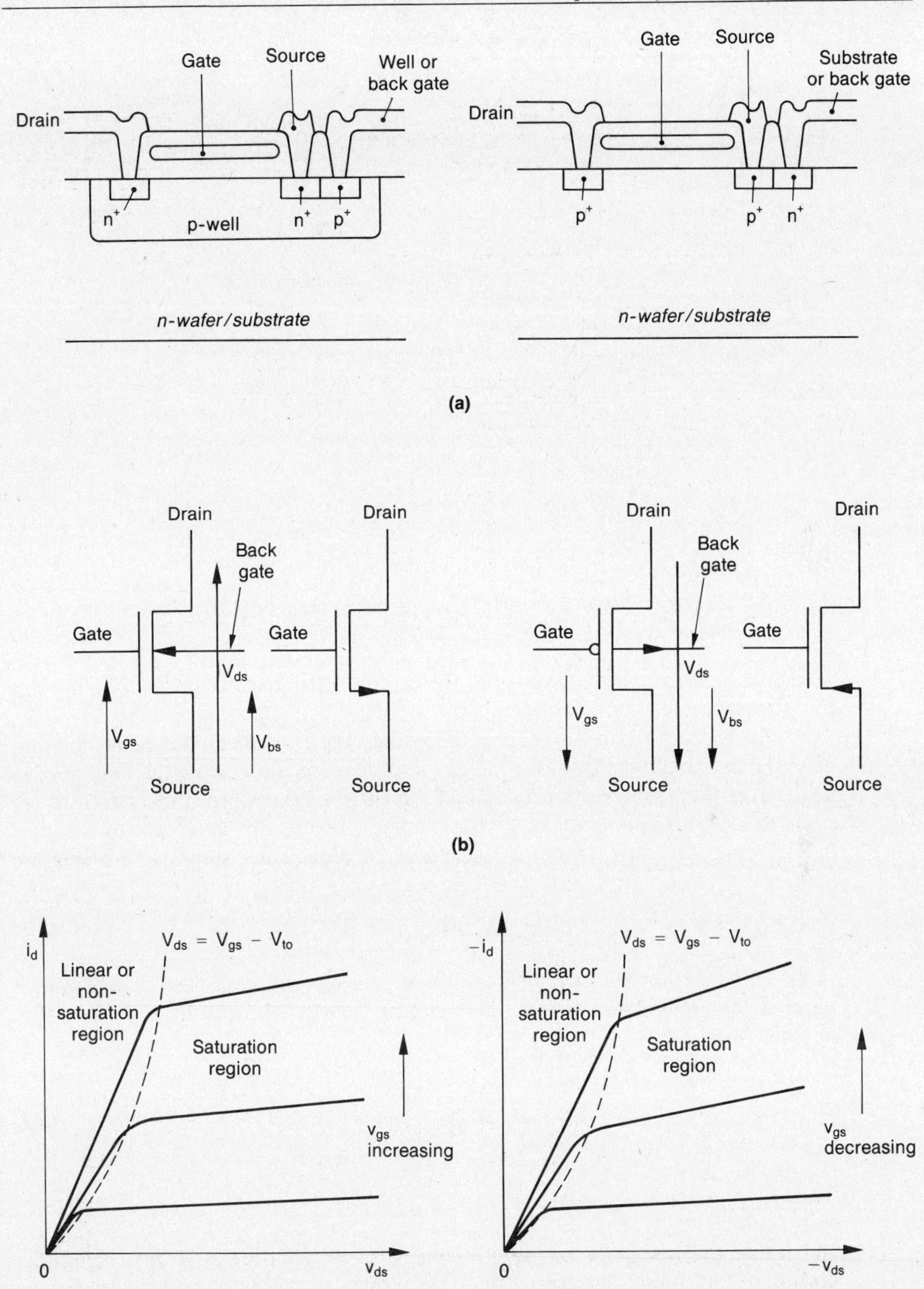

Figure 3.4 Construction, symbol and static characteristics for n- and p-channel enhancement MOS transistors

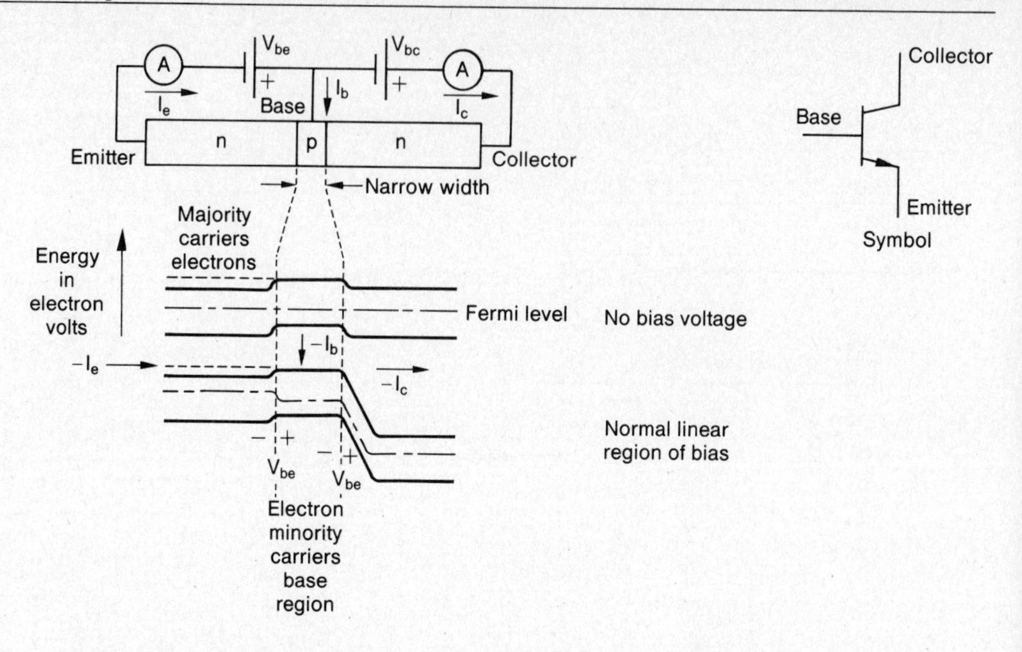

Figure 3.5 Energy diagrams showing the operation of an NPN bipolar transistor

withstand higher levels of dissipation. Hence, it is often made larger in area, but this is not necessary for transistor operation.

In the linear region of operation, the emitter base diode, being forward biased, injects electrons from the emitter into the base. Because of the large reverse field across the base collector diode and the narrow base width, these electrons are swept to the positive potential collector region. Consequently, in a correctly designed bipolar transistor, little current is lost out of the base. Thus, summing currents, we have,

$$I_e = I_b + I_c \approx I_c \tag{3.1}$$

If we define the normal incremental current gain α_N (common base current gain—input is the emitter and output the collector current with the base the common terminal) as,

$$\alpha_N = \left.\frac{\Delta I_c}{\Delta I_e}\right|_{V_{cb} = \text{constant}} \tag{3.2}$$

where $0 < \alpha_N < 1$

and realize that the collector current consists of two portions (the first being that portion of the emitter injected current that reaches the collector and the second the leakage current I_{cbo} of the base collector diode—it is reverse biased), we can write the collector current as,

$$I_c = \alpha_N I_E + I_{cbo} \tag{3.3}$$

Since the emitter current is a junction diode forward current, then

$$I_e = I_{beo}(e^{\frac{qV_{be}}{kT}} - 1) \qquad \textbf{(3.4)}$$

where I_{beo} = the emitter base diode leakage current
V_{be} = the forward voltage across the diode
q = the electron charge
T = the absolute temperature in °K and
k = Boltzman's constant

Substituting into equation 3.3, we obtain,

$$I_c = \alpha_N I_{beo}(e^{\frac{qV_{be}}{kT}} - 1) + I_{cbo} \qquad \textbf{(3.5)}$$

As explained previously, there was no reason why the right-hand side of the transistor in Figure 3.5 was called the collector; as the device is symmetrical. Consequently, if we interchange the collector and emitter, that is, use the collector as the emitter and the emitter as the collector (called the inverted or reverse mode of operation instead of the normal or forward), then our equation would become,

$$I_e = I_{beo} + \alpha_I I_{cbo}(e^{\frac{qV_{cb}}{kT}} - 1) \qquad \textbf{(3.6)}$$

Thus, if we are to allow for the fact that a transistor can function in the normal or inverted mode, the complete equations describing its characteristics are,

$$I_c = \alpha_N I_{beo}(e^{\frac{qV_{be}}{kT}} - 1(+ I_{cbo}(e^{\frac{qV_{cb}}{kT}} - 1) \qquad \textbf{(3.7)}$$

and

$$I_e = I_{beo}(e^{\frac{qV_{be}}{kT}} - 1) + \alpha_I I_{cbo}(e^{\frac{qV_{cb}}{kT}} - 1) \qquad \textbf{(3.8)}$$

These equations can be easily checked: when either of the diodes are reverse biased, the exponential terms tend to unity, giving equations 3.5 and 3.6. These equations are known after Ebers and Moll, the people who first derived them (Ebers and Moll, 1954), and they are fundamental to the understanding and design of bipolar transistor circuits. They are the basis of most analog computer-aided design program models for bipolar transistors.

Before considering these equations further, we will examine several other simple concepts relating to the bipolar transistor. Figure 3.6 shows the symbol and static characteristics of the two types of bipolar transistors, namely npn and pnp. Like the complementary MOS transistors, the theory developed for the npn type applies equally to the pnp except that all voltages and current flows need to be reversed. The characteristics shown are for the common emitter configuration. If equations 3.1 and 3.3 are combined, we obtain,

$$I_c = \left(\frac{\alpha_N}{1 - \alpha_N}\right) I_b + \frac{I_{cbo}}{(1 - \alpha_N)}$$

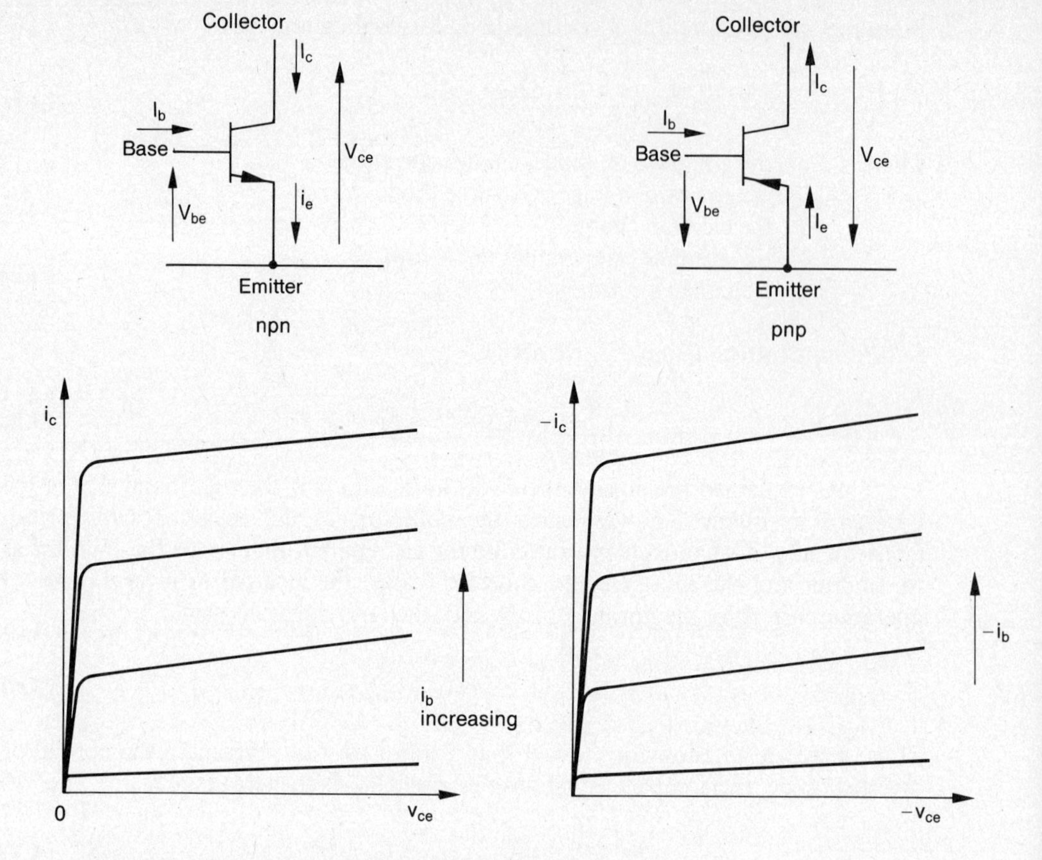

Figure 3.6 Static characteristic of bipolar transistors

Defining the common emitter incremental current gain as,

$$\beta_N = \left.\frac{\Delta I_c}{\Delta I_b}\right|_{V_{cb} = \text{constant}}$$

we discover that,

$$I_c = \beta I_b + I_{ceo} \tag{3.9}$$

where

$$\beta_N = \frac{\alpha_N}{1 - \alpha_N}$$

and

$$I_{ceo} = \frac{I_{cbo}}{(1 - \alpha_N)} = (\beta_N + 1)I_{cbo} \tag{3.10}$$

Returning to equation 3.5, if we define the transconductance of a transistor as,

$$g_m = \left.\frac{\Delta I_c}{\Delta V_{be}}\right|_{V_{ce} = \text{constant}} \tag{3.11}$$

we can show that,

$$g_m = \frac{qI_c}{kT} \tag{3.12}$$

which at 300° K gives,

$$g_m \approx 40\ I_c \tag{3.13}$$

where I_c is the DC collector current flowing in the transistor.

So far we have only considered the operation of the transistor in one of three possible regions of operation, that is, the linear region where one diode was forward biased and the other was reverse biased. Which diode is which depends on whether the transistor was in the normal or inverted mode. The other two regions are where:

1. both diodes are reverse biased while the transistor is nonconducting and said to be "cut off" (equivalent to an open switch); and,
2. both diodes are forward biased and the transistor is saturated or in the "on" region (equivalent to a closed switch).

We can easily see the three regions by examining the circuit and characteristics given in Figure 3.7. On the static characteristics, we can draw the load line (as described by equation 3.14), derived by summing the voltage across the series load resistor and transistor and by equating it to the supply voltage V_{cc}.

$$i_c = \frac{-1}{R_L}V_{ce} + \frac{V_{cc}}{R_L} \tag{3.14}$$

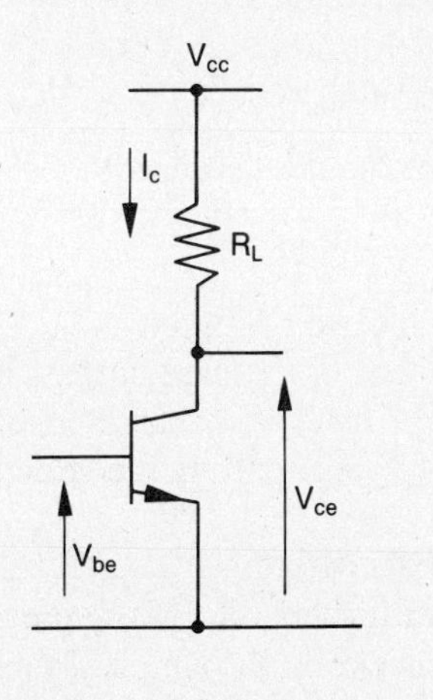

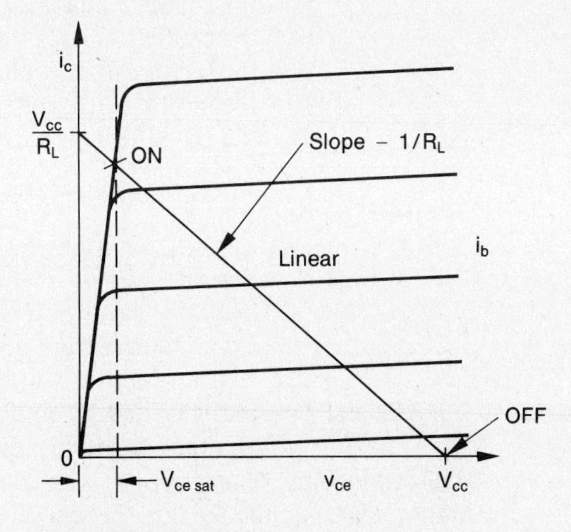

Figure 3.7 Three regions of operation for a bipolar transistor

The load line has a slope of $-1/R_L$ and cuts the Y axis at V_{cc}/R_L. Simple calculations show it must cut the X axis at V_{cc}. The linear region of operation is shown. The "off" region is where the base current is near zero. While V_{be} may not be negative, because of the exponential characteristics of the diode (equation 3.4), the current is so low it approximates zero. The saturation or "on" region is where the load line cuts the vertical portion of the transistor characteristic. Here the collector current is less than β_N times the base current (there is excess base current), so that the collector emitter saturation voltage $V_{ce\,sat}$ becomes 100 or 200 millivolts. Since the base emitter voltage is typically 700 millivolts, because of the large base current, both diodes are forward biased.

It is important to note that the Ebers and Moll equations (3.7 and 3.8) cover all three regions of operation and are therefore a global model.

3.5 Advanced bipolar transistor models

Although the Ebers–Moll model is a very powerful one and all DC large signal models are based on it, there are a number of limitations. Two variations on this model (Getrev, 1976) are the injection and the transport version. They differ only in the choice of reference currents, but the SPICE model employs the second approach. To this model, ohmic bulk resistances r_b', r_c' and r_e', diffusion and junction capacities, base width modulation and charge storage effects must be added to make a more complete model. Figure 3.8 shows the Gummel–Poon model (Gummel and Poon, 1970) used in the analog simulation program SPICE.

For DC modeling of a bipolar transistor, some 14 parameters are important:

β_F or β_N—the common emitter forward current gain
β_R or β_I—the common emitter reverse current gain
I_s—the saturation leakage current
R_b—the base bulk resistance
R_c—the collector bulk resistance
R_e—the emitter bulk resistance
V_A—the forward Early voltage
V_B—the inverse Early voltage
I_K—the forward high knee current for the I_c versus $V_{b'e'}$ characteristic where the slope changes from $\frac{q}{kT}$ to $\frac{q}{2kT}$
I_{KR}—as for I_K but in the reverse or inverse operation
C_2 and N_e—describe the nonideal component of the base current at very low currents as shown up on the I_c versus $V_{b'e'}$ characteristic when the slope changes from q/kT to $q/N_e kT$, cutting the vertical axis at a value greater than I_s, namely $C_2 I_s$
C_4 and N_c—as for C_2 and N_e but for inverse operation.

We can see the need of these parameters if we compare actual transconductance characteristics with those predicted by the simple Ebers and Moll model. Plotting the log of the collector current versus the actual base emitter voltage, equation 3.7 predicts a straight line of slope q/kT (Figure 3.9(a)). In

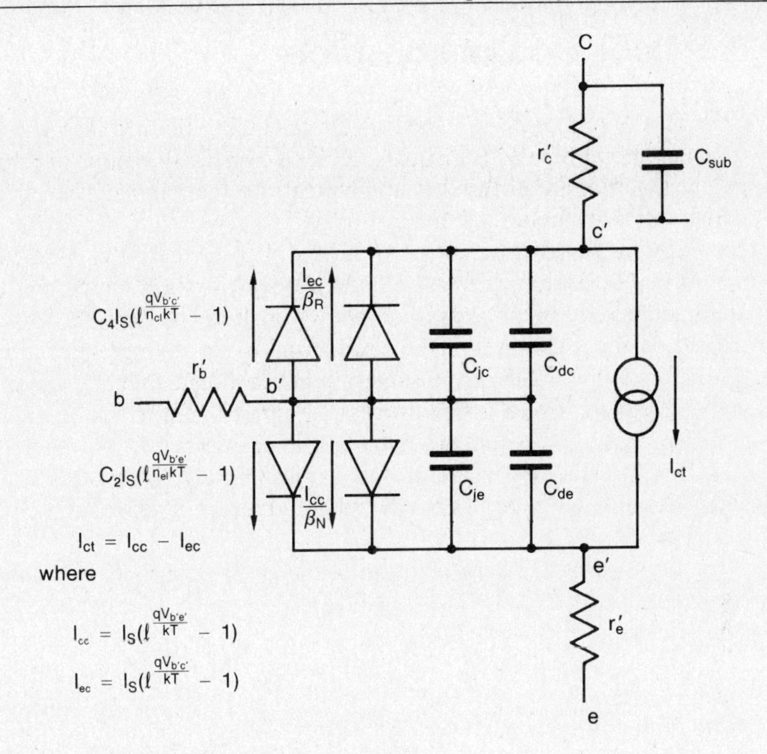

Figure 3.8 SPICE model for an npn transistor

reality, practical transistors deviate from this at both very large and very small currents (as illustrated in Figure 3.9(b)). The figure also shows how some of the additional parameters, needed to define a model of a practical bipolar transistor, had come about (Getreu, 1976). Further information on the determination of these parameters for a practical transistor will be given in Chapter 11.

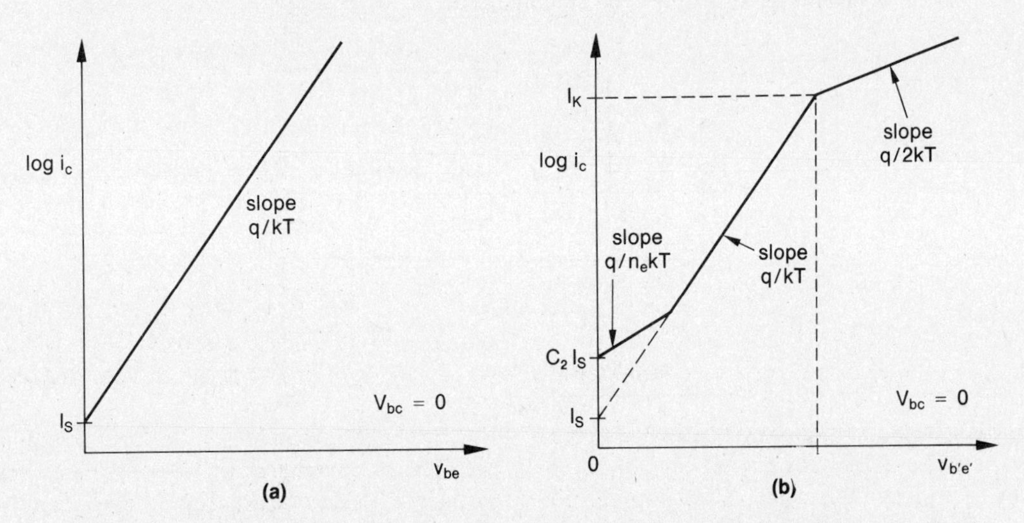

Figure 3.9 Predicted characteristics from Ebers and Moll equations (a) and actual characteristics of a practical transistor (b)

3.6 Practical bipolar transistors

With all the processes discussed so far (including the nMOS process), it is possible to make lateral bipolar transistors. Here two diffused regions of opposite type to the parent material are diffused or implanted into the parent material. In principle, the parent material could be the substrate or a well but, since the parent material becomes the base region of the transistor, it must carry all base currents. Further, for normal operation, the base emitter diode is forward biased so that there must be voltages present of polarity that allow junctions to become forward biased with respect to the substrate. These limitations make it impractical for lateral bipolar transistors to be made in the substrate material and they are therefore restricted to the well regions. For a p-well process, an npn transistor can be made or a pnp in an n-well process. Twin tub processes allow both types to be made. Thus with bulk CMOS processes, both lateral and vertical bipolar transistors are possible, while with SOS only lateral types can be made. Figure 3.10 shows these transistors for the p-well process.

For normal linear operation of the lateral transistor, the collector base diode is reverse biased so that Q_R is cut off. However, Q_L is active and detracts from the performance of the lateral transistor.

As discussed earlier, the performance of the bipolar transistor depends very much on the width of the base region. In the case of the vertical transistor, the base width is the difference between the depth of the p-well and n^+ diffusions. Using the typical figures given in Table 3.2 and assuming a nominal well depth of 4 μm, the base width for the metal gate process is 1.5 μm and for the silicon gate n-well, 3.5 μm. Consequently, the degrading effect of the vertical transistor will be much less for this n-well process than the metal gate. On the other hand, the gain of the vertical transistor on the n-well process will probably be so low that such a device cannot be included in designs on its own merits. In practice, as source drain junction depths are decreased, well depths are also made shallower so that the gain of the vertical transistor may even be enhanced.

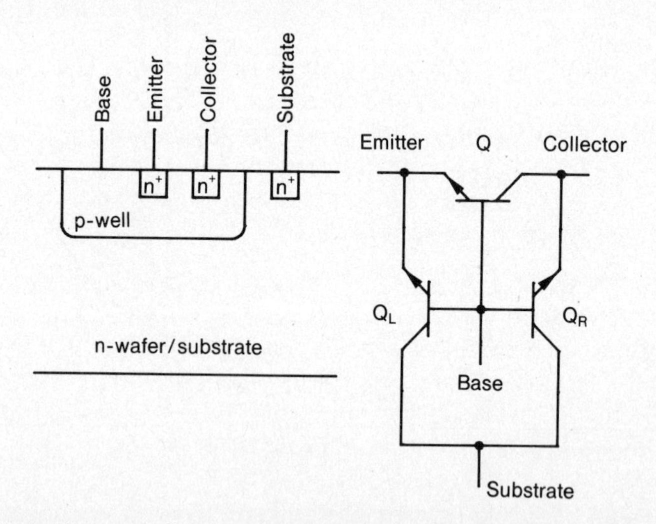

Figure 3.10 Lateral and vertical npn transistors in a p-well

When making a lateral bipolar transistor, three simple rules must be followed. First, since the emitter will emit in all directions, it is necessary to totally surround the emitter with a collector to ensure a high collection efficiency. This is illustrated in Figure 3.11. Second, the adverse effect of the vertical bipolar transistor can be minimized if the emitter area of the lateral transistor is made as small as possible, that is, of minimum dimensions. It can be shown (Colclaser, 1980) that the common emitter current gain β_N and unity gain frequency f_T, with and without the parasitic vertical transistor present, are as follows:

1. *Current gain.*
   No parasitic present,

$$\beta_N = \frac{2L_p^2}{W_L^2}$$

Parasitic present,

$$\beta_N = \frac{A_e}{A_p}\left(1 + \frac{2L_p^2}{W_L^2}\right) \tag{3.15}$$

2. *Frequency at which unity gain.*
   No parasitic present,

$$f_T = \frac{D_p}{\pi W_L^2}$$

Parasitic present,

$$f_T = \frac{D_p}{\pi W_L^2} \frac{A_e}{(A_e + A_p)} \frac{\left(1 + \dfrac{A_p}{A_e}\dfrac{W_L}{W_p}\right)}{\left(1 + \dfrac{A_p}{A_e}\dfrac{W_p}{W_L}\right)} \tag{3.16}$$

where L_p = the minority carrier diffusion length in the well region
W_L = the base width of the lateral transistor
W_p = the base width of the parasitic transistor
D_p = the carrier diffusivity
A_e = the sidewall emitter area or lateral emitter area
A_p = the plan area or parasitic emitter area

The effect of the parasitic can be minimized by maximizing the ratio A_e/A_p.

Finally, to improve the gain of the lateral device, the base width should be made the minimum dimension allowed by the process, that is, the closest that two diffusion regions can be placed together. For the lambda based CMOS rules given in Figure 3.13 (p. 58), this is 3 lambda (3λ). Where this spacing is greater than the minimum width of a polysilicon line (as is the case here, namely, 2λ), an alternative approach is to make a square MOS transistor (Figure 3.11) and then apply a zero or reverse bias source gate voltage (Vittoz, 1983). The second approach often produces a transistor that does not have a minimum dimensions emitter and it

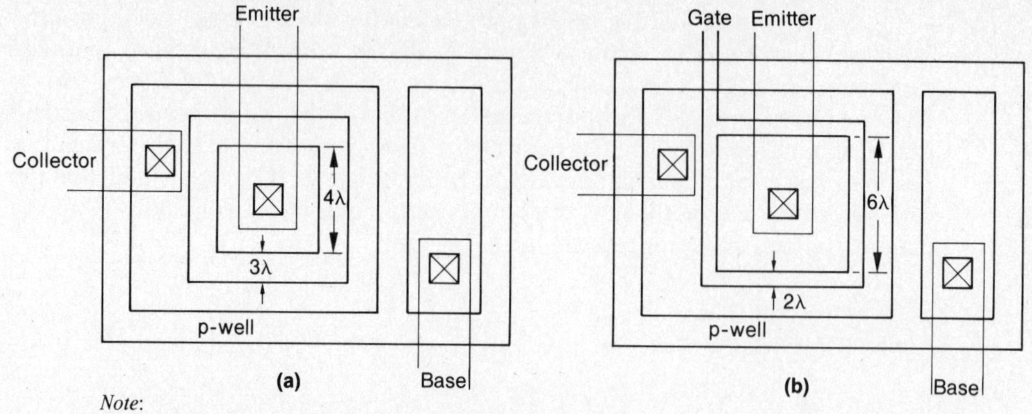

Note:
The second method (b) can often achieve a smaller base width

Figure 3.11 Two methods of making a lateral bipolar transistor

occupies slightly more silicon area even though the base width is narrower. However, its characteristics are often superior.

The bipolar transistor offers two advantages in analog work. First, the transconductance (equations 3.12 and 3.13) is much higher than that of a MOS transistor and, second, the base emitter characteristics (or v_{be} versus i_e or i_c) are usually well matched. This is important for current mirrors (Chapter 4) and band gap references (Chapter 6). The fact that these circuit configurations have common bases allows all the transistors to be in the same well.

Tables 3.3 and 3.4 provide measured information on lateral and vertical transistors for several processes. Minimum dimension emitters were used in every case.

Table 3.3 Typical parameters for conventional lateral (Figure 3.11(a)) and vertical transistors

	p-well process		*n-well process*		
Parameter	Lat. npn	Vert. npn	Lat. pnp	Vert. pnp	*Unit*
β_F	20	1000	2.5	150	
β_R	3	1.4	0.25	0.02	
I_S	$1.4.10^{-13}$	$1.5.10^{-14}$	8.10^{-13}	$1.6.10^{-13}$	amp
R_b	1500	700	2000	900	ohm
R_c	20	—	30	—	ohm
R_e	2	—	3	—	ohm
V_A	13	36	7	30	volt
V_B	1.5	5.5	15	10	volt
I_K	6.10^{-5}	$1.0.10^{-4}$	$1.0.10^{-4}$	$2.4.10^{-4}$	amp
I_{KR}	1.10^{-5}	$5.0.10^{-5}$	$1.0.10^{-4}$	$1.4.10^{-6}$	amp
C_2	$6.8.10^5$	$6.7.10^4$	$7.5.10^3$	1.10^4	
N_e	5.1	2.5	2.8	2.15	
C_4	$2.1.10^5$	$7.7.10^2$	$1.5.10^3$	$1.7.10^2$	
N_c	4.6	2.2	4.0	2.18	

Note:
Emitter areas 4λ by 4λ.

Table 3.4 Typical parameters for a MOS type, n-well, lateral PNP transistor (Figure 3.11(b))

Parameter	*MOS type lateral pnp transistor*	*Units*
β_F	40	
β_R	40	
I_S	2.10^{-13}	amp
R_b	2000	ohm
R_c	40	ohm
R_e	5	ohm
V_A	18	volt
V_B	15	volt
I_K	$1.6.10^{-4}$	amp
I_{KR}	5.10^{-6}	amp
C_L	6.10^{4}	
N_e	2.8	
C_4	—	
N_c	—	

Note:
Emitter area 6λ by 6λ.

3.7 Latch up

Although the more complex CMOS process allows novel bipolar transistors, these very transistors can cause problems. Consider two bipolar transistors connected as shown in Figure 3.12(a). The emitter currents flowing in both transistors must be identical and equal to the sum of the two collector currents. Thus,

$$\begin{aligned} I_e &= I_{cp} + I_{cn} \\ &= \alpha_p I_e + I_p + \alpha_n I_e + I_n \end{aligned}$$

where α_p and α_n = the common base current gains
I_p and I_n = the collector base leakage currents

A rearrangement results in the following:

$$I_e = \frac{I_p + I_n}{1 - (\alpha_p + \alpha_n)} \tag{3.17}$$

At very low currents, transistors normally have low current gains. Thus $(\alpha_p + \alpha_n)$ tends to zero and the emitter currents flowing are only small leakage currents. However, if through injected base currents, radiation or an increase in temperature (leakage currents double approximately every 8°C temperature rise) the sum of the two current gains reaches unity, then the emitter current becomes very large and is limited only by the bulk resistances of the devices. This is called *latch up*. The transistors can only be turned off be removing the power rail as well as the offending cause of the increase in current gains.

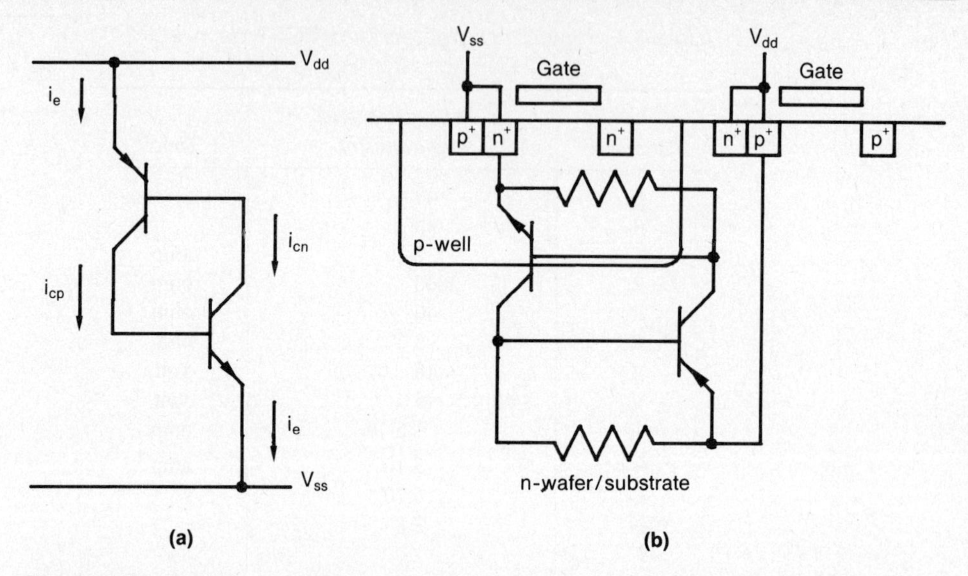

Figure 3.12 Thyristor operation or latch up operation in a bulk CMOS process

Considerable effort is being made to analyze and reduce this problem (Hall et al., 1985; Yilmaz, 1985; Chen and Wu, 1985; Lipman, 1982). First, however, we need to ask how this npn/pnp transistor configuration comes about in CMOS. Figure 3.12(b) shows how it is possible in a bulk CMOS process. The pnp transistor is formed as follows: the source of a p-channel MOS transistor acts as the emitter, the n-wafer substrate, the base and the p-well becomes the collector. For the npn transistor, the source of an n-channel MOS transistor is the emitter, the p-well and the base and the substrate is the collector.

Latch up can be eliminated (Lipman, 1982) by, first, reducing the transistor gains through increasing base width. Second, by decreasing the size of two inherent resistors which are in parallel with the base emitter diodes (Figure 3.11(b)), the chance of turning on the transistors is considerably reduced. Third, most foundries have layout rules to assist in reducing latch up, particularly the placement of supply rail contacts to a well. For example, a typical p-well process statement may be that no point in a well should be at a distance greater than typically 30λ from an ohmic contact while contacts should be as close as possible to transistor sources. Finally, placing p^+ guard rings around p-wells and n^+ guard rings around n-wells is another solution, but it significantly increases the silicon area required.

3.8 Passive components

Because CMOS involves additional processing steps, a larger range of passive components is available than with the simpler nMOS process. There are two noticeable additions. The first is the high well resistance of typically 5 kohm/square, providing a way of making high value resistors, occuping small silicon areas. The second is that we can now have forward biased diodes. Perhaps we should not say diode, for the diode must be in a well, and that immediately implies, for bulk CMOS processes, a vertical bipolar transistor whose collector is the substrate.

A warning in the use of polysilicon resistors may be helpful here. Since both p- and n-implants are used in the process, polysilicon in different parts of the chip can finish up with different levels of doping and hence sheet resistance. Table 3.5 provides details of the sheet resistances and capacities per unit area or periphery for several CMOS processes.

Table 3.5 Typical parameters for resistance and capacity for various bulk CMOS processes

	Metal Gate p-well			*Si Gate p-well*			*Si Gate n-well*			
Parameter	Min.	Typ.	Max.	Min.	Typ.	Max.	Min.	Typ.	Max.	*Unit*
n^+ sheet resistance	5	10	15	10		30	25	30	35	ohm/square
p^+ sheet resistance	30	60	100	30		30	70	80	90	ohm/square
Well sheet resistance				3		4	4		5	kilo-ohm/square
Polysilicon sheet resistance				15		30	20	25	30	ohm/square
Metal sheet resistance								30		milli-ohm/square
Gate capacity		1.0		4.5		5.0		0.7		10^{-4} pF/μm^2
Source drain capacity										
n^+ area	1.5		3.0	3.0		3.5	1.0		1.3	10^{-4} pF/μm^2
n^+ perimeter	4.0		8.0		5.0			6.0		10^{-4} pF/μm
p^+ area	0.6		1.5		2.0			2.0		10^{-4} pF/μm^2
p^+ perimeter	2.0		4.0		3.0			4.0		10^{-4} pF/μm
Well capacity										
Area					0.77			0.33		10^{-4} pF/μm^2
Perimeter					3.3			5.0		10^{-4} pF/μm
Poly to substrate					0.5			0.43		10^{-4} pF/μm^2
Poly to metal					0.4			0.41		10^{-4} pF/μm^2
Metal to substrate					0.23			0.21		10^{-4} pF/μm^2

Note:
For junction capacities, reverse bias is 2 volts.

3.9 CMOS layout rules

Each of the various CMOS processes has a different set of layout rules and mask requirements. Since the purpose of this book is to concentrate on circuit design aspects, a single set of portable lambda based rules is presented in Figure 3.13, Table 3.6 and Plate 2. The rules do not specify a thin oxide layer but concentrate on p^+ and n^+ diffusion regions, for these are the areas of immediate concern to the designer. Consequently, when designs are completed, the design information must be translated into a format suitable for mask manufacture. Before a simple conversion program can be written, specific mask details must be known. For a simple silicon gate p-well process, details for all masks, except the thin oxide (Mask 2) and p^+ (Mask 4), are in the required form. The thin oxide is formed by performing the logical OR operation of the layout of the n^+ and p^+ region. In the case of the p^+ mask, the implanted p region must be 2λ greater in each direction than the p^+ layout to ensure that, in spite of process and alignment inaccuracies, a

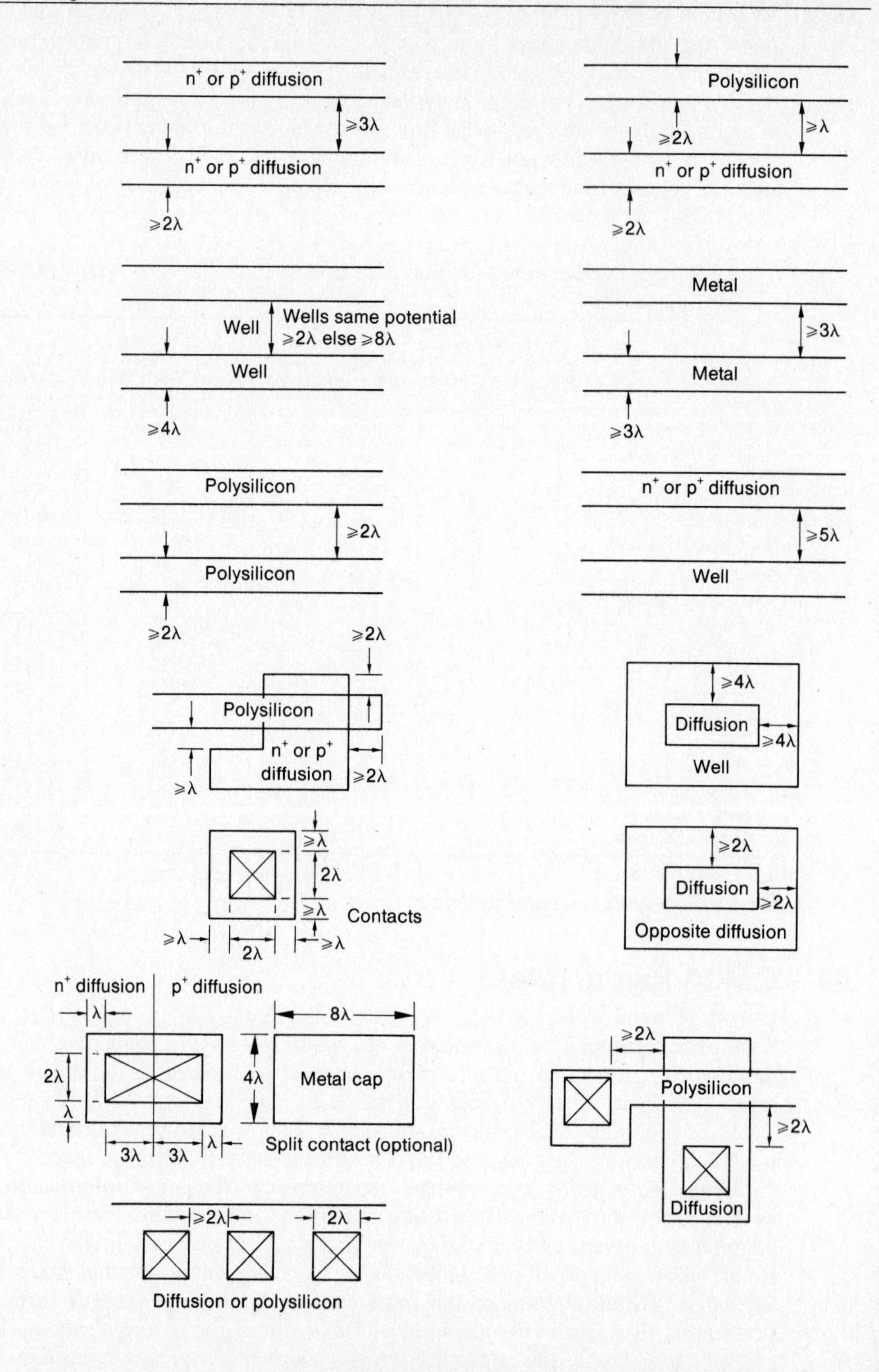

Figure 3.13 Portable CMOS lambda based layout rules

Table 3.6 Colors for the various CMOS layers

CMOS layers	*Colors*
Well material	Brown
n^+ diffusion	Green
p^+ diffusion	Yellow
Polysilicon	Red
Contact	Black
Metal	Blue
Glaze	Orange (or brown)
Double layer metal or polysilicon	Purple

p-channel MOS (or p-contact) is always formed (Figure 3.14). Thus the p^+ mask is generated by increasing the area of every p^+ region in the layout by adding to each dimension typically 4λ.

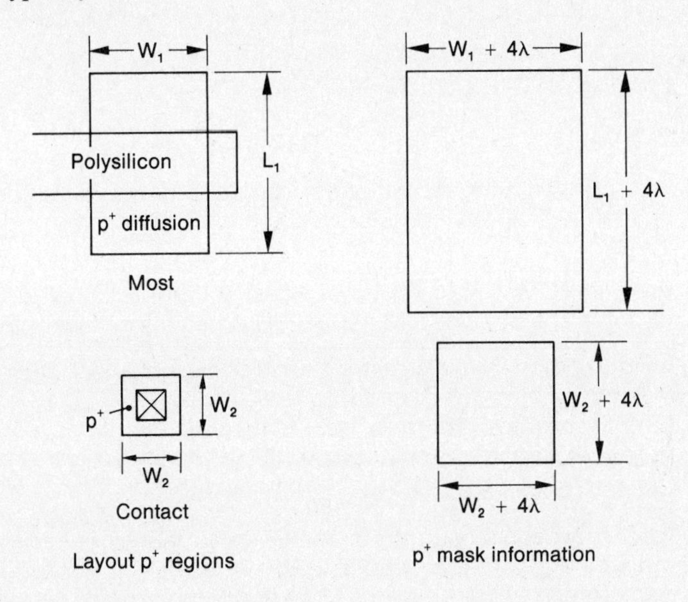

Figure 3.14 Generation of p^+ information—Mask 4 in the p-well process

3.10 Exercises

1 Using the parameters given in Table 3.2, derive the transfer characteristic of a CMOS inverter (Figure 3.2) for transistors having a W/L ratio of 2. What is the incremental gain of the circuit when self-biased (biased so input voltage equals output voltage)?

2 Figure 3.15 shows circuits of one and two transistor analog switches. Using parameters for the silicon gate p-well process of Table 3.2, plot the switch resistance against applied gate voltage for all three circuits. Assume an input voltage of 1 to 4 volts and switching signal voltages over the range of 0 to 5 volts. All transistors are of unity aspect ratio.

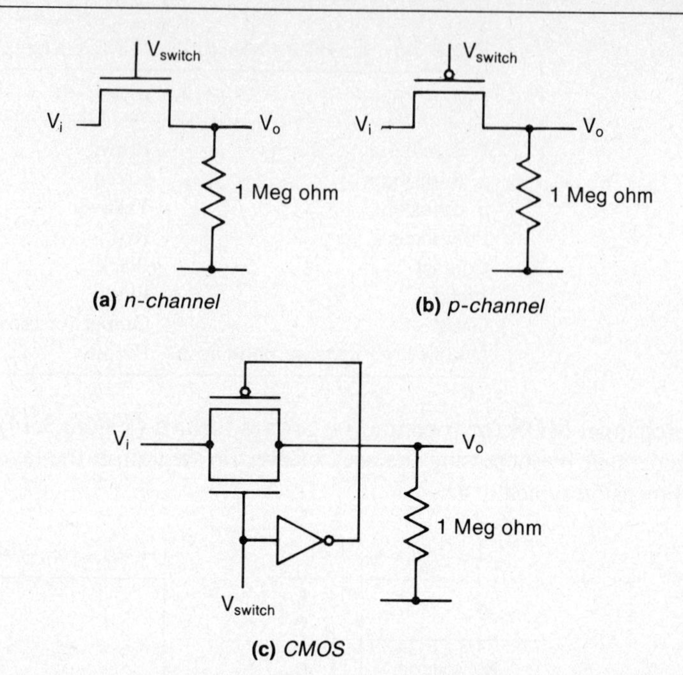

Figure 3.15

3 Use equation 3.4 to show that the intrinsic small signal resistance looking into the emitter of a bipolar transistor is 260 ohm at 300°K and for an emitter current of 100 μA.

4 Using the CMOS layout rules of Figure 3.13 for a p-well process, complete the layout for the two types of lateral NPN transistors, as shown in Figure 3.10.

5 Refer to Figure 3.16. How would you modify your layout derived in Question 4 to achieve:

(a) two transistors of equal emitter area; and
(b) two transistors of emitter area 3:1,
without appreciably increasing silicon area?

6 Latch up occurs because of the parasitic four layer (thyristor/SCR) device. In many industrial applications, such a device is required. Using the SPICE parameters of Tables 3.2 and 3.3 for the silicon gate p-well process, calculate the input current required to turn "on" the current mirror thyristor circuit of Figure 3.17.

7 The DC collector leakage current for an open base transistor is given by equations 3.9 and 3.10. Using simple logic show that,

$$I_c = (\beta + 1)I_{cbo}$$

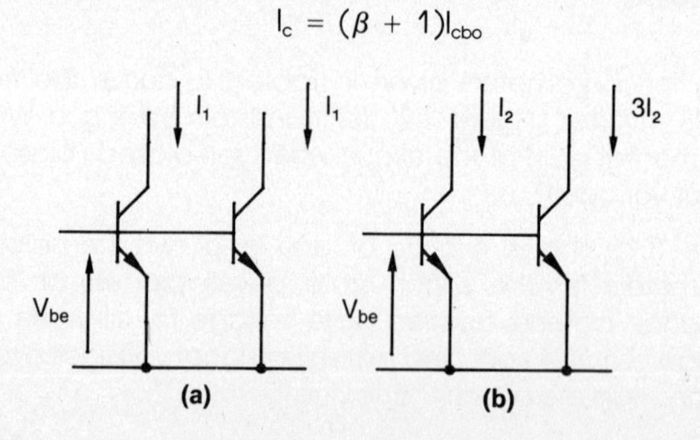

Figure 3.16

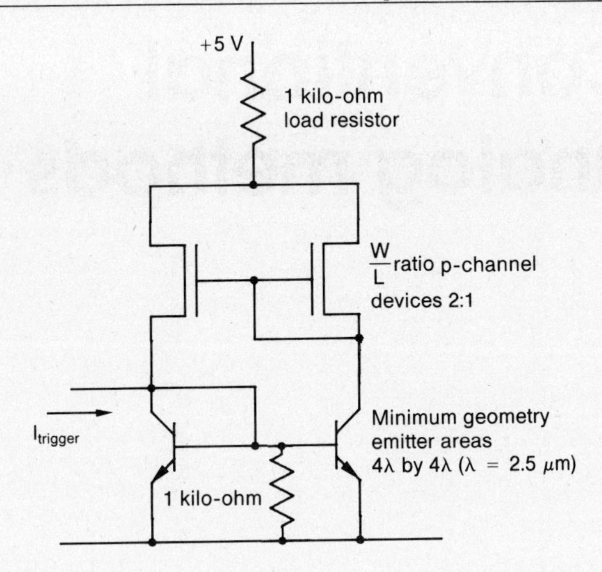

Figure 3.17

3.11 References

Chen, M. J. and Wu, C. Y. (1982), "A Substrate Orientated Model for Determining the Substrate Spreading Resistance in Bulk CMOS Latch Up Paths and its Application in Holding Current Predicting", *Solid State Electronics*, Vol. 28, pp. 855–66.

Colclaser. R. A. (1980), *Microelectronics, Process and Device Design*, John Wiley, New York.

Ebers, J. J. and Moll, J. L. (1954), "Large Signal Behaviour of Junction Transistors", *Proceedings IRE*, Vol. 42, pp. 1761–72.

Getrev, I. (1976), "Modeling the Bipolar Transistor", *Tektronics*, Oregon.

Grisworld, T. W. (1982), "Portable Design Rules for Bulk CMOS", *VLSI Design*, Sept.–Oct., pp. 62–7.

Gummel, H. K. and Poon, H. C. (1970), "An Integral Charge Control Model of Bipolar Transistors," *Bell Systems Tech-Journal*, Vol. 49, pp. 827–52.

Hall, J. E., Seitchik, J. A., Arledge, L. A. and Yang, P. (1985), "An Improved Circuit Model for CMOS Latch Up", *Trans IEEE Electron Device Letters*, Vol. EDL–6, pp. 320–2.

Joint Microelectronic Research Centre (1985), *CMOS Layout Rules*, University of NSW, Sydney.

Lipman, J. (1981), "A CMOS Implementation of an Introductory VLSI Design Course", *VLSI Design*, Fourth Quarter, pp. 56–8.

— (1982), "Latchup Protection in P Well Bulk CMOS Circuits", *VLSI Design*, Vol. 3, p. 30.

Natural Sciences and Engineering Research Council of Canada (1985), *Guide to the Integrated Circuit Implementation Services of the Canadian Microelectronics Corporation*, Natural Sciences and Engineering Research Council of Canada, Queen's University, Kingston, Canada.

Vittoz, E. A. (1983), "MOS Transistors Operated in the Lateral Bipolar Mode and their Application to CMOS Technology", *Solid State Circuits*, IEEE Journal, Vol. SC–18, pp. 273–9.

Weste, N. and Eshraghian, K. (1985), *Principles of CMOS VLSI: A Systems Perspective*, Addison-Wesley, Reading, Mass.

Yilmaz, H. (1985), "Cell Geometry Effect on ICT Latch Up", *IEEE Electron Device Letters*, Vol. EDL–6, pp. 419–21.

4 Conventional analog methods

4.1 The operational amplifier

In the world of discrete devices, the champion analog device is the operational amplifier (Graeme et al, 1971). Much of what we have come to expect of a typical operational amplifier is based on what is achievable from a bipolar rather than a MOS process. For example, we may expect the following characteristics:

1. Open loop gain (differential) $> 10^5$.
2. Gain bandwidth product $>$ 1 MHz.
3. Input resistance $> 10^5$ ohm.
4. Output resistance (open loop) $< 10^2$ ohm.
5. Offset voltage $<$ 1 mV.
6. Dynamic range $>$ 80 percent of supply rail.
7. Common mode rejection ratio $>$ 60 dB.

The amplifier itself normally consists of four stages, as shown in Figure 4.1. The input stage is a high gain differential amplifier followed by a differential to single ended converter. It is possible to take the output from only one side of the differential stage but this halves the gain. Next, additional voltage gain circuits are included and these may also contain components to ensure that the operational amplifier is stable. The final part is a power stage so that the amplifier can deliver large load currents.

In bipolar technology, the circuits most used to build up the various stages of the amplifier are current mirrors. The same technique can be employed with MOS circuits (two current mirrors are shown in Figure 4.2). Consider the basic circuit. Since both transistors have the same gate source voltage, the currents, which when both transistors are in the saturation region of operation, are governed by equation 2.13. Consequently, the current ratio I_o/I_i is determined by the aspect ratio of the transistors.

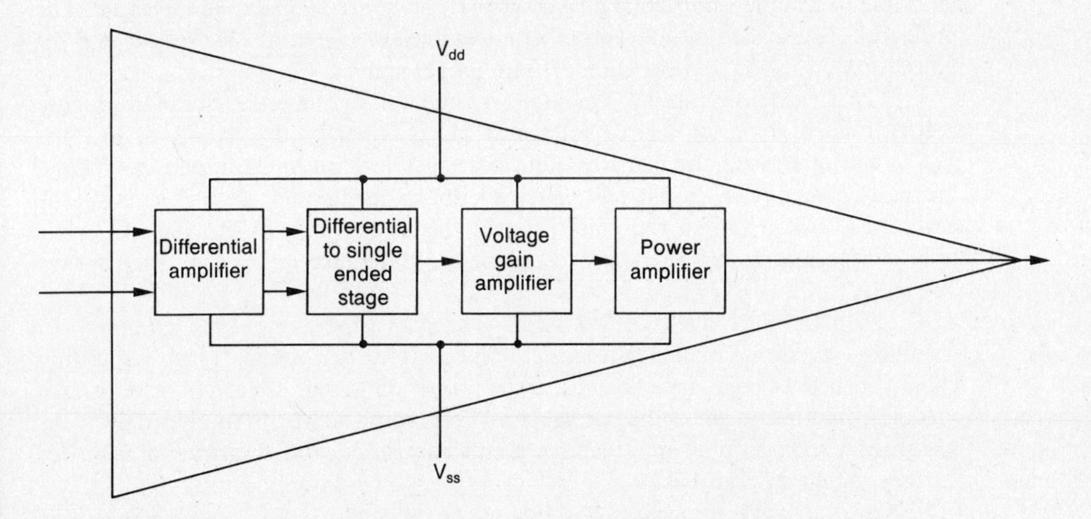

Figure 4.1 Basic stages of a conventional operational amplifier

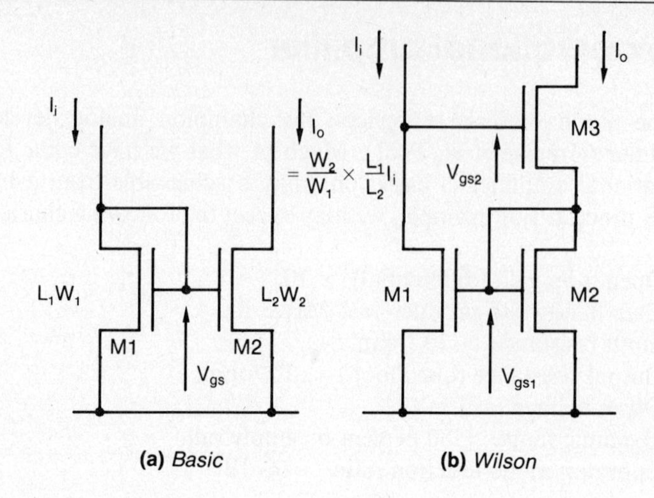

Figure 4.2 Current mirror

$$(V_{gs} - V_{to})^2 = \frac{2I_i}{\mu C_{ox}} \times \frac{L_1}{W_1} = \frac{2I_o}{\mu C_{ox}} \times \frac{L_2}{W_1}$$

or

$$\frac{I_o}{I_i} = \frac{L_1}{W_1} \times \frac{W_2}{L_2} \tag{4.1}$$

For identical transistors, the ratio is unity, that is, the output current mirrors the input current. With the simple two transistor circuit, it would be unusual for the two drain source voltages to be equal. Consequently, the current ratio has an error term due to channel shortening effects when drain source voltages are unequal. The problem is aggravated in that a small drain source voltage across M2 puts it into the nonsaturation region and equation 2.13 no longer applies.

This situation can be overcome by the three transistor circuit in Figure 4.2(b). Firstly, the drain source voltage of M1 is typically $2V_{gs}$, while for M2 it is V_{gs}, which minimizes the error by limiting the difference in drain source voltage. Secondly, the circuit employs negative feedback techniques to help control the current ratio. The current ratio I_o/I_i is set by the aspect ratios of M1 and M2, while a large aspect ratio for M3 helps to reduce the difference in the two drain source voltages.

A better circuit for nMOS is a four transistor type which employs the modified inverter circuit (Haskard, 1983). Here the near ideal saturation characteristics of the depletion transistor, operating with negative gate source voltages, minimizes any voltage changes. (See Figure 4.3(a).) An alternative four transistor circuit employ only enhancement transistors, and therefore suitable for CMOS, shown in Figure 4.3(b).)

With CMOS, both complementary MOS transistors and lateral bipolar types are available. These two factors make current mirrors in CMOS technology as

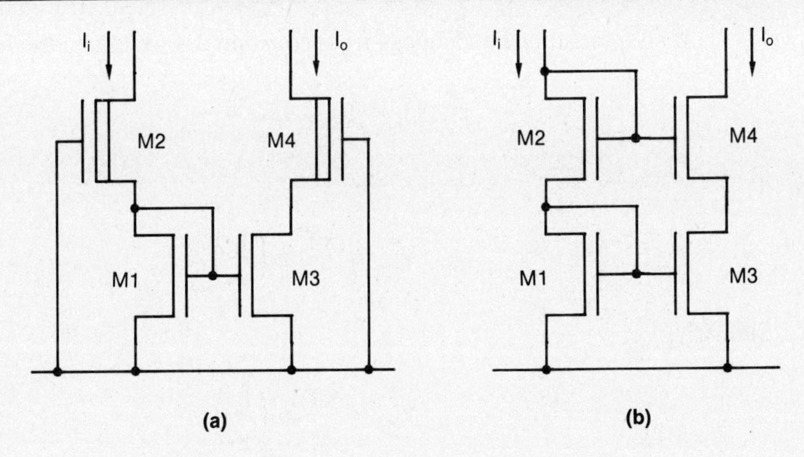

Figure 4.3 Four transistor current mirrors

powerful as in bipolar. Complementary transistors allow current mirror sources and sinks while the bipolar transistor mirror allows accurate low current mirroring. This is made possible by the repeatability of the base emitter characteristics of a junction diode. However, to allow for second order effects, accurate current ratios can only be achieved by placing identical transistors in parallel. This tends to limit the size of the ratio to the range $\frac{1}{8}:8$. Noninteger ratios can be achieved with less accuracy by using an emitter resistor (as shown in Figure 4.4). Ratios of less than or greater than unity can be achieved simply by changing the resistor from one side of the circuit to the other.

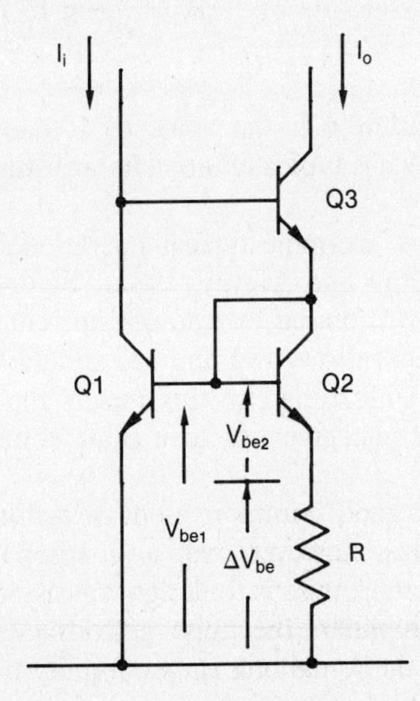

Figure 4.4 Bipolar transistor Wilson current mirror with a resistor to determine current ratio

If all transistors are identical, from equation 3.8 we obtain the following:

$$I_i \approx I_{e1} \approx I_{beo}\ell^{\frac{qV_{be1}}{kT}}$$

or

$$V_{be1} = \frac{kT}{q} \ln\left(\frac{I_1}{I_{beo}}\right) \tag{4.2}$$

Similarly,

$$V_{be2} = \frac{kT}{q} \ln\left(\frac{I_o}{I_{beo}}\right) \tag{4.3}$$

and since

$$\Delta V_{be} = V_{be1} - V_{be2}$$

$$\Delta V_{be} = \frac{kT}{q} \ln\left(\frac{I_i}{I_o}\right)$$

However,

$$\Delta V_{be} = I_o R$$

hence,

$$R = \frac{kT}{qI_o} \ln\left(\frac{I_i}{I_o}\right) \tag{4.4}$$

Using equation 4.4, the value of R can be determined for any required currents. Because V_{be} is typically 0.6 volts, resistor values are usually not excessively large.

We will now examine typical operational amplifier building block circuits (Tsividis, 1978; Gray and Meyer, 1982). Figure 4.5 shows a nMOS differential amplifier (M1 to M4) biased by a simple current mirror (M5 to M7). To achieve a high gain, the aspect ratio of M1 and M2 should be large, while that of M3 and M4 should be small. Unfortunately, this means the DC voltages V_o and V_r tend to approach V_{ss}, and can give rise to a poor common mode voltage range for the amplifier.

To achieve good common mode rejection, M5 should appear as an ideal current source. This, however, will also upset the common mode input voltage range. A typical compromise is to design transistor M5 of near unity aspect ratio.

For any amplifier, the noise performance is determined primarily by the input stage, as all the remaining stages amplify both the noise of the amplifier and the input signal. In Section 2.7 we saw that the noise of a transistor could be minimized by using transistors which are large in area. Consequently, if amplifier noise is important, then M1, M2, M3 and M4 should all be made large in area.

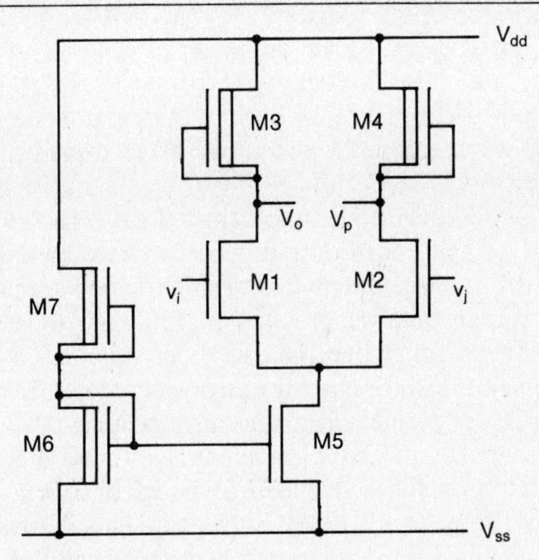

Figure 4.5 Simple differential amplifier

Because the design of the differential stage is a delicate balance between DC output voltages and gain, it is difficult to achieve high gains in a single nMOS stage. A typical gain may be in the range 20 to 40. Techniques such as load compensation (Toy, 1979) can be applied to improve the gain, but as a general rule the added complexity is not warranted. Differential stages can be cascaded to increase gain.

A CMOS differential stage is similar to the nMOS amplifier (shown in Figure 4.5) except that M3 and M4 can be p-channel MOS transistors, while M7 is normally a resistor or when it is an n-well process, several serial n-channel enhancement transistors. Due to n-channel transistors having high back gate voltage sensitivity, in many p-well processes the differential amplifier load resistors are n-channel transistors so that their sources can be at zero substrate potential (Smarandoiu et al., 1978; Gregorian and Nicholson, 1979) Such a circuit is shown in Figure 4.6. The p-well is tied to the source of M1 and M2 to eliminate back gate effects.

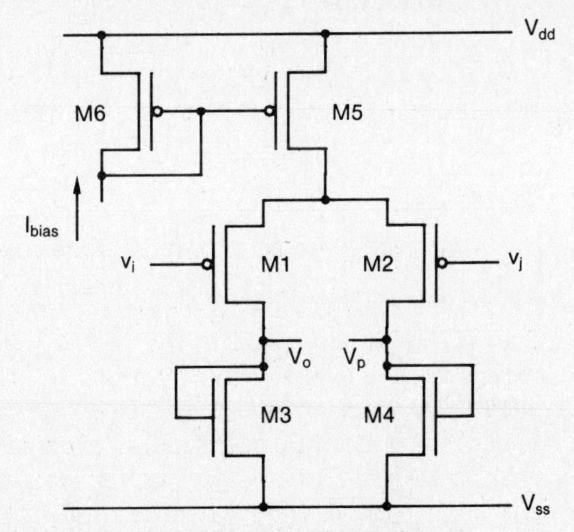

Figure 4.6 A p-well process differential amplifier design

The input stage of an amplifier is generally responsible for the circuit's overall noise and offset voltage. Unfortunately, MOS transistor input stages can result in a poor performance for both of these parameters. One solution (Gustafsson et al., 1984) is to replace the two input MOS transistors by Darlington connected bipolar transistors.

Following the input stage (or stages), the differential output must be converted to a single ended output. This can be achieved by using a source follower (Figure 4.7(a)) or a true differential to single ended amplifier (Figure 4.7(b)).

The source follower provides a significant loss in gain and is only used in simple amplifiers. Notice that the true differential to single ended amplifier employs an enhancement transistor current mirror as the load. In CMOS transistors, M1 and M2 would be p- or n-channel enhancement types.

Alternatively, in CMOS, it is possible to use bipolar techniques and connect the differential amplifier active load as a current mirror to directly provide a single ended output (Stone et al., 1984). Such a circuit is illustrated in Figure 4.8.

Any gain stages that now follow are usually of the two transistor inverter form with the differential to single ended stage providing a DC level shift to bias the inverter stages into their linear region. The frequency response of these amplifiers in nMOS can be improved by using a split load transistor (Tsividis, 1977).

In nMOS, the final or output stage is often similar in configuration to that of the digital super buffer or super inverter. The transistors operate in a class AB mode. Figure 4.9 shows a typical circuit. With CMOS, there are at least two options. A MOS complementary output stage could be employed, as shown in Figure 4.10(a). Alternatively, one or more of these transistors are designed as bipolar devices to capitalize on the high transconductance to deliver large load currents (see Figure 4.10(b)).

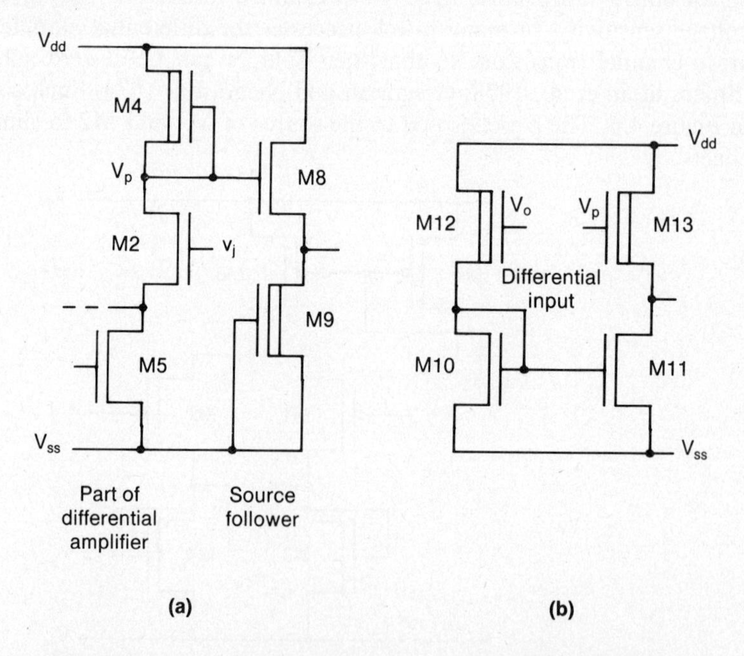

Figure 4.7 Differential to single output circuits for the nMOS process

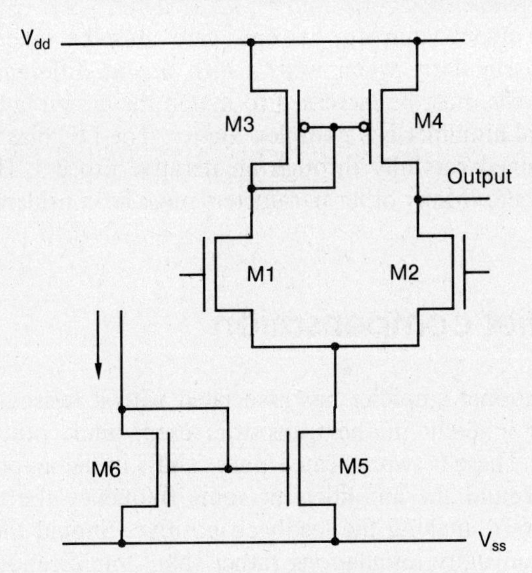

Figure 4.8 Differential input amplifier with single ended output

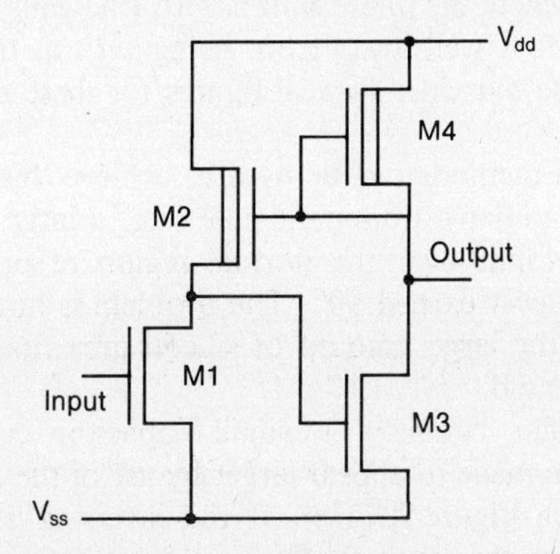

Figure 4.9 Power output stage for an nMOS process

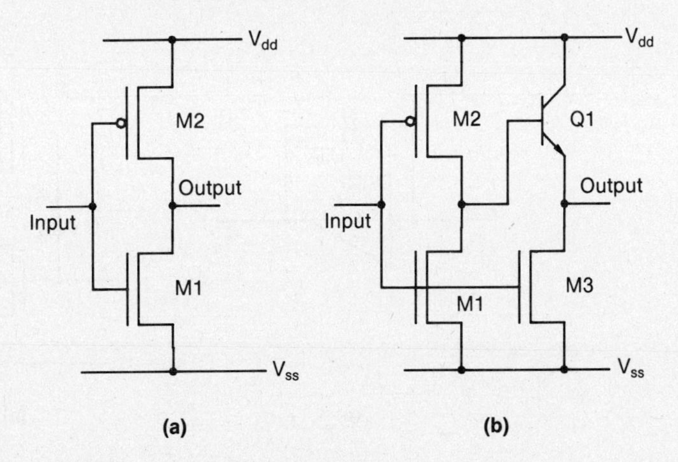

Figure 4.10 Typical CMOS process output stages

To avoid latch up, extreme care must be taken when laying out CMOS circuits, particularly when wells which are at different potentials. Also, spacing between wells must be increased to match the design layout rules. The design of an operational amplifier is a complex matter. The DC bias levels and stage gains must be determined carefully through an iterative process. However, the problem does not stop there. Many other parameters must be considered, especially stability.

4.2 Amplifier compensation

The operational amplifier has associated with it resistive and reactive components. Some are inherent in the transistors used, while others, due to the layout, are parasitics. These networks cause phase shifts in the amplifier. If negative feedback is applied around the amplifier, at some frequency the total loop phase shift may become 360°, making the feedback positive. Should the gain at this frequency be greater than unity, oscillations rather than amplification will result. An important part of amplifier design is to ensure that the gain of the amplifier is less than unity at the frequency where the phase shift is zero (the gain margin) or, conversely, that the loop phase shift is well away from being zero at the frequency where the gain is unity (the phase margin). Typical figures for these margins are 10 dB for gain and 45° for phase.

Various methods can be used to achieve this, the simplest being to give the operational amplifier a dominant pole (e.g. a large shunt capacitor to ground) so that the phase shift over the normal region of operation (voltage gains from 1 upwards) can never exceed 90°. The problem is how to achieve this compensation without using the large amount of silicon area that would be taken by the shunt capacitor to ground.

In bipolar circuitry, a small capacitor correctly positioned around an amplifier can be made to appear larger by use of the Miller feedback effect.

Consider Figure 4.11(a). If the inverting amplifier is ideal, all the input current flows into the capacitor C.

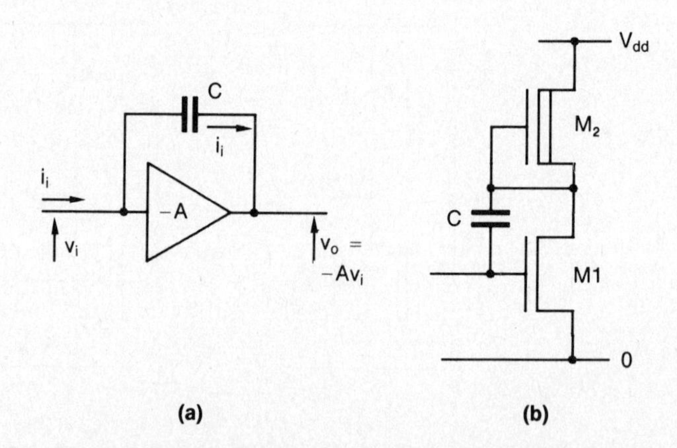

Figure 4.11 Capacitance amplification using the Miller effect

Thus,

$$i_i = \frac{v_i - (-Av_i)}{\frac{1}{j\omega C}}$$

$$= v_i(A + 1)j\omega C$$

where $1/j\omega C$ is the reactance of the capacitor at frequency ω.

Now,

$$Z_i = \frac{v_i}{i_i} = \frac{1}{j\omega C(A + 1)}$$

so the apparent input capacity to the circuit resulting from feedback is,

$$C_i = (A + 1)C \qquad \textbf{(4.5)}$$

When we use this technique with a simple MOS inverter (Figure 4.11(b)), a problem occurs because we have both feed forward and feedback, since a signal can flow either way through the capacitor. We therefore not only generate our required dominant pole, but also a zero (Tsividis, 1978).

Consider the equivalent circuit for the inverter circuit of Figure 4.11(b) (as shown in Figure 4.12). Generator i_i is the current from the previous stage, R_i and R_o the input and output resistance respectively of the inverter, g_m its transconductance and C the feedback capacitor.

Summing currents at nodes 1 and 2 gives:

$$v_i = \frac{i_i + v_o sC}{sC + 1/R_i}$$

and

$$v_i = \frac{(sC + 1/R_o)v_o}{(sC - g_m)}$$

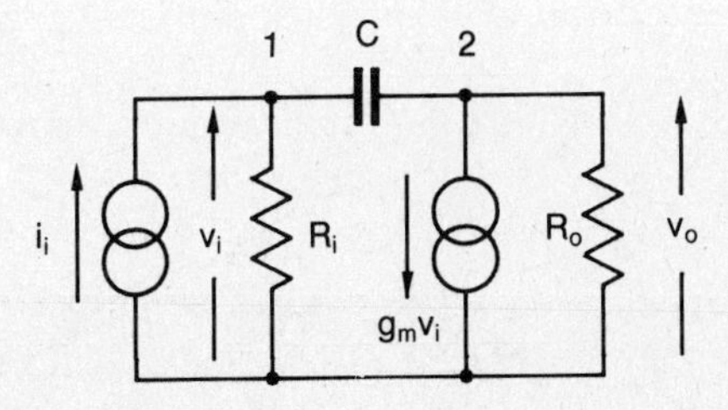

Figure 4.12 Equivalent inverter circuit for the Miller capacitor multiplier circuit of Figure 4.11(b)

Combining and solving gives:

$$\frac{v_o}{i_i} = \frac{(s - g_m/C)}{\left[\frac{1}{R_i} + \frac{1}{R_o} + g_m\right]\left[s + \frac{1}{CR_o + CR_i + CR_oR_ig_m}\right]} \tag{4.6}$$

Thus we not only have a pole at frequency,

$$\omega_p \simeq \frac{1}{(CR_oR_ig_m)} \tag{4.7}$$

but a zero at frequency,

$$\omega_z = \frac{g_m}{C} \tag{4.8}$$

Bode plots for the amplifier are shown in Figure 4.13. With bipolar technology, the transistor conductance is high so that the frequency ω_z is well outside the range of useful gains. For low transconductance MOS transistors, this is not the case and the feed forward signal must be eliminated in some way. Two methods of doing this are to add a resistor in series with the capacitor (McGreary, 1983), or to add a buffer source follower (as shown in Figure 4.14). Repeating the previous calculations for these cases (Figure 4.14), we find for case (a) that,

$$\frac{v_o}{i_i} = \frac{(1 - g_mR)\left[s - \frac{g_m}{C(1 - g_mR)}\right]}{\left[\frac{1}{R_i} + \frac{1}{R_o} + g_m + \frac{R}{R_oR_i}\right]\left[s + \frac{1}{(R_oC + CR_i + Cg_mR_oR_i + CR)}\right]} \tag{4.9}$$

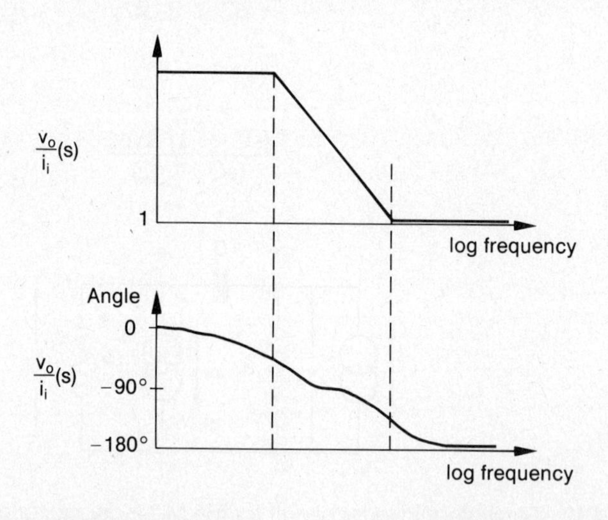

Figure 4.13 Bode plots of equation 4.6

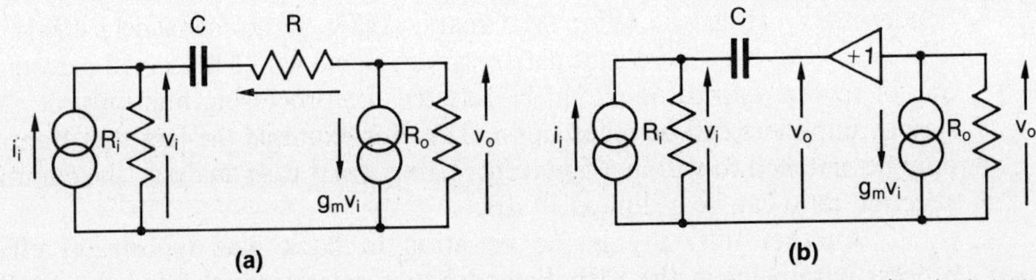

Figure 4.14 Two methods of preventing the effects of feedforward causing instability

giving a pole at the frequency of approximately $1/Cg_mR_oR_i$ but a zero at $g_m/C(1 - g_mR)$. By controlling the value of the series resistor R, we can shift the zero out to a frequency beyond the unity gain frequency. The value of R is noncritical as it does not matter how far the zero is shifted out beyond the unity gain frequency. Figure 4.15(a) illustrates this method for CMOS.

Examining case (b), we discover the zero has disappeared altogether for,

$$\frac{v_o}{i_i} = \frac{-g_mR_o}{C(1 + g_mR_o)(s + 1/R_iC(1 + g_mR_o))} \tag{4.10}$$

There is still the pole at the same approximate frequency of $1/Cg_mR_oR_i$. Figure 4.15(b) illustrates this second method for nMOS.

4.3 Further difficulties in conventional amplifier design

The design of an operational amplifier is a complex procedure, particularly for the nMOS process. Many of the processes available for multiproject chips have wide spreads in parameters, so extra safeguards must be inbuilt to ensure that the amplifier remains stable under all operating conditions. Thus, if a process has parameters defined to ±10 percent, complex operational amplifier designs could be undertaken (Young, 1979; Tsividis et al., 1976; Senderowicz et al., 1978; Fry 1969;

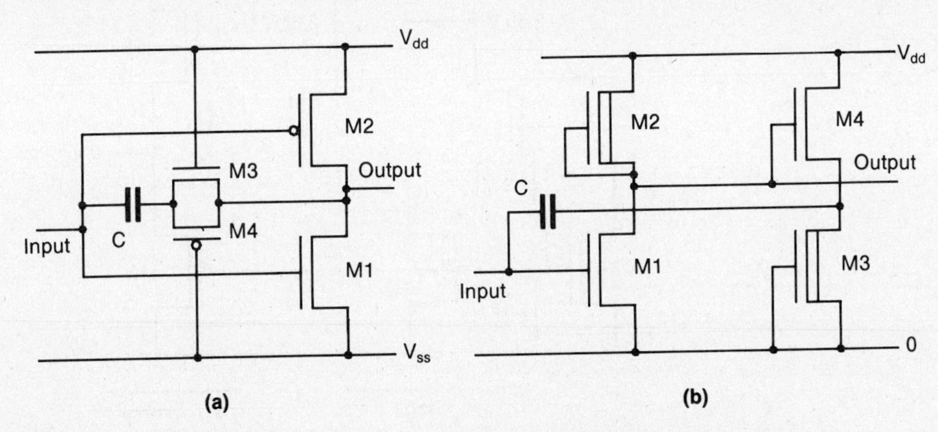

Figure 4.15 Typical MOS operational amplifier compensation circuits

Senderowicz, Huggins, 1982; McCreary, 1983; Krummenacher, 1981; Wu and Mavor, 1983), but where parameters vary as much as +100 to −50 percent, and any one of several fabricators could be selected for processing multiproject chips, only simple amplifiers can be relied upon. Take for example the fact that transistors can only be matched to within ±5 percent. Using worst case analysis, the common mode rejection ratio can be as low as 14 dB.

Another difficulty is the variation in back gate (substrate) effects. The process parameter in the SPICE model that determines this is gamma. Because it can vary by a factor of 4:1, the operating point for any transistor whose source is not at the substrate potential can vary considerably. Thus amplifier parameters, such as dynamic range and common mode input voltage range, can vary by large amounts from process to process as well as from batch to batch. Using current mirrors driven from a common source, or other feedback techniques between stages, some stabilizing effect can be obtained, but experience suggests that well characterized high gain operational amplifiers cannot be made with reasonable yields using many of the nMOS fabrication processes.

4.4 Commencing an amplifier design

The approach suggested in this book is that "hand" calculations should be undertaken to test an idea before full computer simulation is carried out (Figure 2.5). Let us apply this approach to the design of the on chip CMOS operational amplifier (as shown in Figure 4.16). Specifications for the amplifier are:

1. open loop gain typically 2400;
2. output stage standing current 60 μA to charge and discharge capacity loads;
3. supply requirements are ±5 volts with current drain less than 100 μA;
4. the n-well silicon gate process (as given in Table 3.2) to be used.

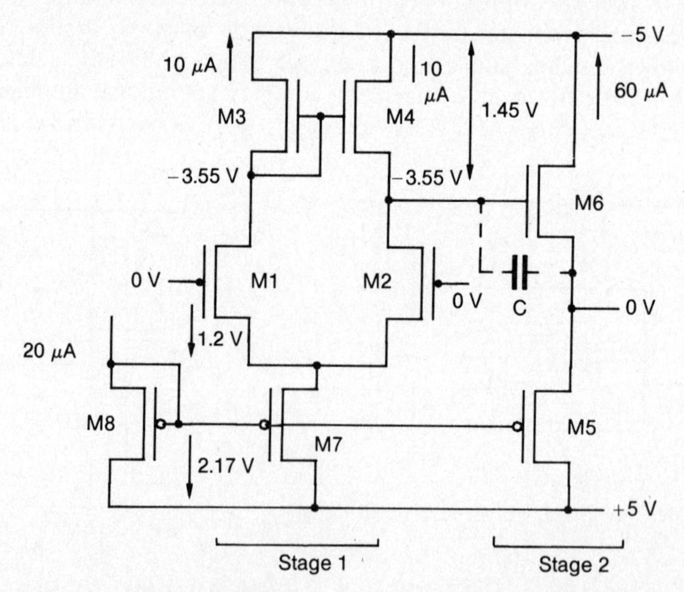

Figure 4.16 On chip CMOS amplifier

When undertaking the design, there are three criteria that must be considered:

1. DC conditions—bias currents, voltages;
2. AC conditions—gain, impedance, stability;
3. layout requirements—area, aspect ratio, power and signal rail positions.

Often there is a conflict between the criteria and a compromise must be reached. For the initial design phase, we will concentrate on the first two.

Using the simple small signal model given in Figure 2.8, an equivalent circuit of the amplifier can be drawn (Figure 4.17). Analysis shows the gain for stage 1 to be,

$$A_{v1} = \frac{v_l}{(v_j - v_i)}$$

$$= \frac{g_{m1}g_{m3}}{(g_{m3} + g_{d1} + g_{d3})(g_{d1} + g_{d3})}$$

and for good devices,

$$g_{m3} \gg g_{d1} + g_{d3}$$

giving,

$$A_{v1} = \frac{g_{m1}}{g_{d1} + g_{d3}} \tag{4.11}$$

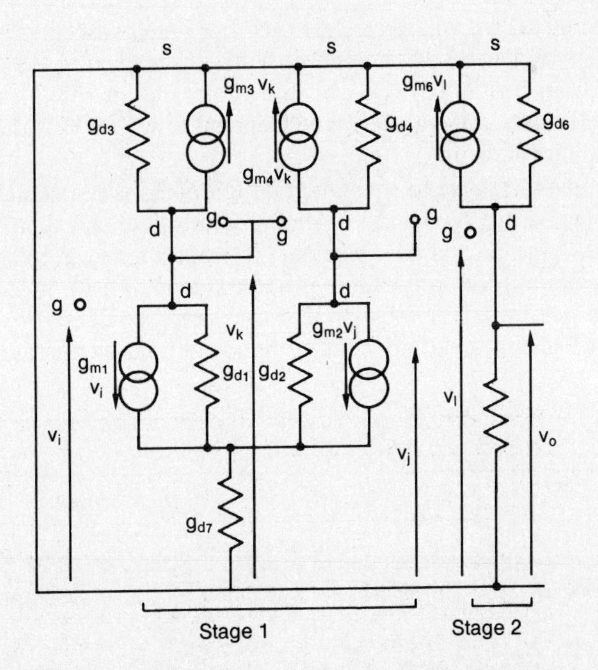

Figure 4.17 Small signal equivalent circuit of the CMOS amplifier

The analysis assumes the M1 and M2 are identical devices as are M3 and M4. Similarly, the stage 2 gain is,

$$A_{v2} = \frac{v_o}{v_1} = \frac{-g_{m6}}{g_{d6} + g_{d5}} \tag{4.12}$$

As a starting point, we will assume that:

1. the drain currents of M1 and M2 are 10 μA;
2. the drain current of M6 is 60 μA; and
3. the overall gain of 2400 will be split between the two stages, say 60 for stage 1 and 40 for stage 2.

Using these currents, the parameters in Table 3.2 together with equations 2.20 and 2.21, approximate values for transistor parameters can be calculated (Table 4.1). Substituting these figures into our two gain equations 4.11 and 4.12 and equating to respective gains of $Av_1 = 60$ and $Av_2 = 40$ gives an aspect ratio W/L for M1, M2 and M6 of 9.

We now need to establish the DC conditions. This can be done using the static model of equation 2.13. Every transistor in the circuit has the substrate or well tied to the appropriate source. As a result, the back gate effects need not be considered. For M6, we have, $W/L = 9$ and $I_d = 60\ \mu A$ giving a gate source voltage of 1.45 volts, and similarly for M1 (or M2) a gate source voltage of −1.21 volts. These voltages can be inserted into Figure 4.16. We now have for M3 equal drain source and gate source voltages of 1.45 volts and $I_d = 10\ \mu A$, giving for M3 and M4 an aspect ratio W/L of 1.48. Transistor M7 has a drain current of 20 μA and a drain source voltage of 3.8 volts. For a unity aspect ratio, a gate source voltage of 2.17 volts is needed. This source voltage is applied to the gate of M5, which must pass a drain current of 60 μA or three times that of M7. Thus an aspect ratio of 3 is needed. Finally, M8 can be made identical to M7 so that the total current taken by the amplifier adds up to 100 μA.

The final part to this initial design phase is the actual transistor dimensions. As explained in Chapter 2, to achieve good noise characteristics, devices should be made as large as possible. Further, to improve device matching thereby reducing offset voltages, the fabrication process should not be pushed to its limits. Again large devices should be used. However, the penalties for using large devices are increased silicon area and reduced bandwidth. If a minimum dimension of 20 μm

Table 4.1 Parameter values for p- and n-channel devices for the silicon gate n-well process assuming $\lambda V_{ds} \ll 1$

	Drain current			
	10 μA		60 μA	
Parameter	n-channel	p-channel	n-channel	p-channel
g_m (μA/V)	$22\sqrt{W/L}$	$13\sqrt{W/L}$	$54\sqrt{W/L}$	$32\sqrt{W/L}$
g_d (μA/V)	0.15	0.5	0.9	3

(or 8λ for λ = 2.5 μm) is used, then resulting transistor dimensions are as given in Table 4.2.

At this point, a full SPICE analysis can be undertaken, then a trial layout to complete device and parasitic capacities. This is followed by a hand calculation to estimate the size of capacity, necessary to achieve stability (under all anticipated loads). Equations 4.6 and 4.7 are used. Finally, full simulation can be undertaken to trim the design to desired performance, including common mode and power supply rejection ratios.

Table 4.2 Transistor dimensions in lambda for the circuit given in Figure 4.16

Transistor	*Width*	*Length*
M1	72	8
M2	72	8
M3	12	8
M4	12	8
M5	24	8
M6	72	8
M7	8	8
M8	8	8

4.5 Typical amplifier designs

In this section we will initially describe two simple nMOS amplifiers that are suitable for many analog applications. A 10 volt supply line is required which could be derived from a 5 volt DC supply using an astable and voltage doubler circuit.

The first circuit (Figure 4.18 and Plate 3) is a unity gain amplifier, capable of supplying large output currents, to charge up capacitors (see Chapter 6) or drive resistor strings (see Chapter 7). It can also be used as an on chip buffer amplifier to drive the outside world. Figure 4.18 is the circuit diagram. It consists of a differential input stage driving a differential to single ended stage. This latter amplifier also acts as an output power stage. The aspect ratios of M7 and M8 can be adjusted to increase or decrease the standing current and hence output drive current. The output is fed back to one side of the differential input, providing 100 percent feedback and a gain of $A/(1 + A)$, where A is the open loop gain. Notice that the input transistors have been made large to keep the noise low.

Plate 3 shows the layout of the amplifier while Table 4.3 provides typical information when the amplifier is made on two different processes.

While many circuits, including filters (Mavor et al., 1981) and voltage references (Haskard, 1983), can be made using a near unity gain amplifier, there are many occasions when the designer wishes to have control of the feedback applied to the amplifier. For this reason, the second example is a simple low gain operational amplifier. The circuit given in Figure 4.19 consists of a differential stage, source follower and inverter power output stage. A shunt capacitor is required to stabilize the amplifier. Typical characteristics for this amplifier are given in Table 4.4.

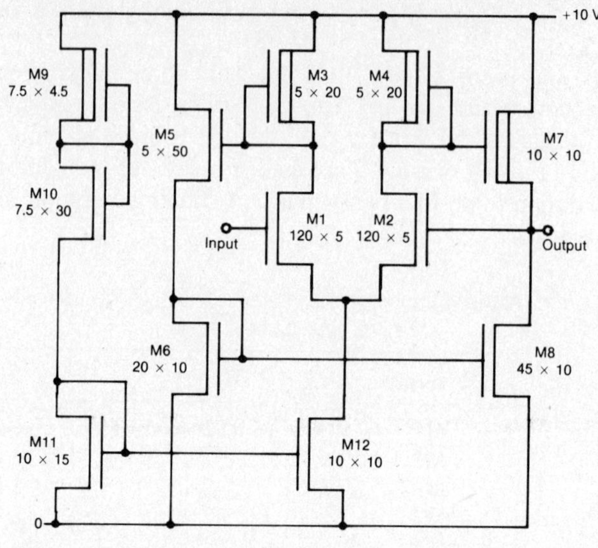

Note:
The dimensions are in microns with width given first.

Figure 4.18 Unity gain buffer amplifier

Table 4.3 Typical no load performance figures for the nMOS unity gain buffer amplifier (V_{dd} = +10 volts)

Parameter	*Manufacturer 2*	*Manufacturer 4*	*Units*
Gain	0.97	0.94	—
Offset	15	160	mV
Dynamic range	2–4.5	2.5–8	volts
Output resistance	1.24	3.5	kilo-ohms
Cut-off frequency	3.5	0.085	MHz
Slew rate	5.0	0.5	V/μsec.
Supply current drain	120	80	μA

Table 4.4 Typical no load characteristics of the low gain operational amplifier with V_{dd} = +10 volts

Parameter	*Manufacturer 2*	*Units*
Open loop gain	60	—
Dynamic range	1–4	volts
Output resistance	75	kilo-ohm
Cut-off frequency	40	kHz
Slew rate	0.6	V/μsec.
Gain as a follower	1.007	—
Current drain	1	mA

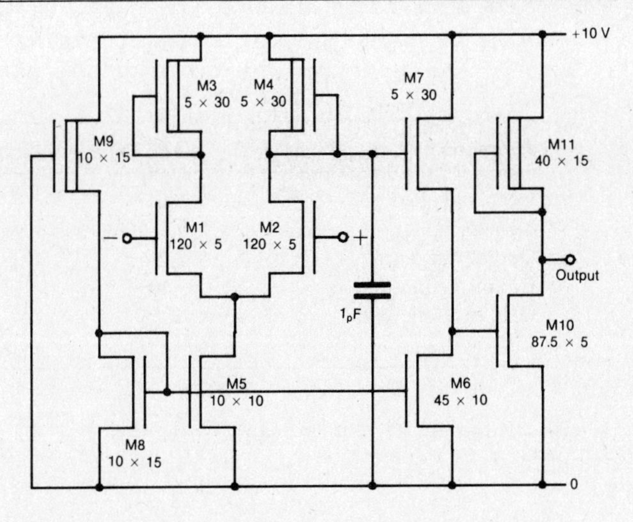

Note:
Dimensions are in microns with width given first.

Figure 4.19 Low gain operational amplifier

The gain when connected with 100 percent negative feedback is greater than unity, as the common mode voltage gain performance is poor.

The remaining two amplifiers are for the CMOS process. Figure 4.20 and Plate 4 show a simple two stage amplifier operating at very low current levels. As each stage is current starved to keep supply currents low, stage gains and impedances are high. Table 4.5 shows typical parameters for this amplifier.

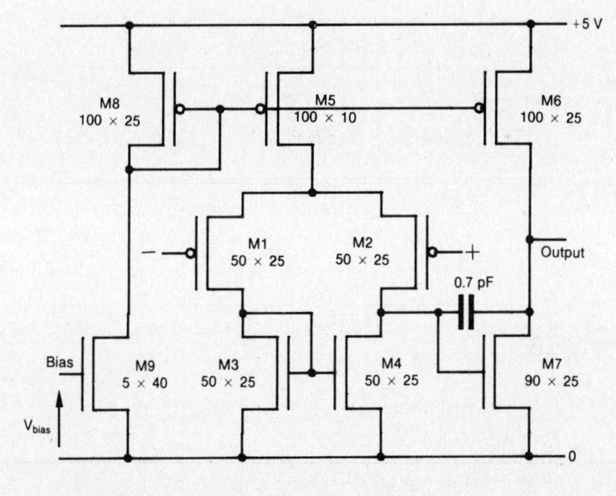

Note:
Device dimensions are given in microns with width first.

Figure 4.20 Low current n-well CMOS amplifier

Table 4.5 Typical no load characteristics for the low power CMOS operational amplifier with V_{dd} = 5 volts

Parameter	*Values*	*Units*
Open loop gain	230 000	—
Dynamic range	0.2–4.3	volts
Output resistance	82	Meg ohm
Cut-off frequency	400	kHz
Gain as a follower	1.00	—
Current drain	0.9	μA

Finally, Figure 4.21, Plate 4 and Table 4.6 give details of an alternative p-well amplifier.

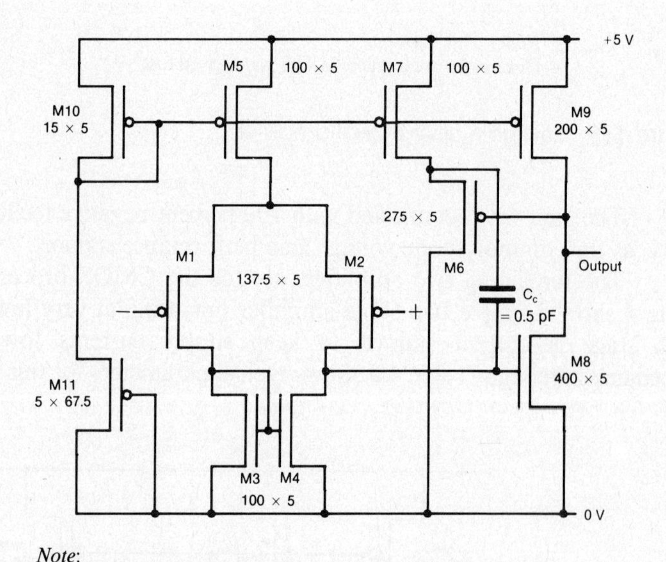

Note:
Device dimensions are given in microns with width given first.

Figure 4.21 Simple p-well amplifier

4.6 Exercises

1 Operational amplifiers are a basic building block for conventional analog circuits. Prepare a list of applications for such an amplifier.

2 Refer to Figure 4.2. Using the simple models derived in Chapter 2, calculate the output currents in terms of the input current when:
(a) all transistors are the same;
(b) transistors M2 and M3 are identical and each consists of three M1 transistors in parallel.

Table 4.6 Typical no load characteristics for a p-well silicon gate CMOS operational amplifier with V_{DD} = 5 volts

Parameter	*Value* Open loop	Closed loop	*Units*
Dynamic range	0.2–4.4	same	volts
Output resistance	80	0.2	kilo-ohm
Cut-off frequency			
C_c = 5 pF	.3	N/A	kHz
C_c = 0.5 pF	7.9	N/A	kHz
Unity gain frequency			
C_c = 5 pF	140	N/A	kHz
C_c = .5 pF	900		kHz
Gain	370	1.00	—
Current drain	55	same	μA
Peak output source current	30	same	μA
Offset voltage	10	same	mV
Slew rate			V/μsec.
C_c = 5 pF	.3	same	mV/μsec.
C_c = .5 pF	70	same	mV/μsec.
THD at 3Vp to p output	2.5	.05	%

3 Refer to Figure 4.22(a). Using identical transistors show for the case of a current mirror an error term for the ratio of output to input current of $2/\beta$ where β is the transistor current gains. What happens to this error term when an enhancement transistor is added? See Figure 4.20(b).

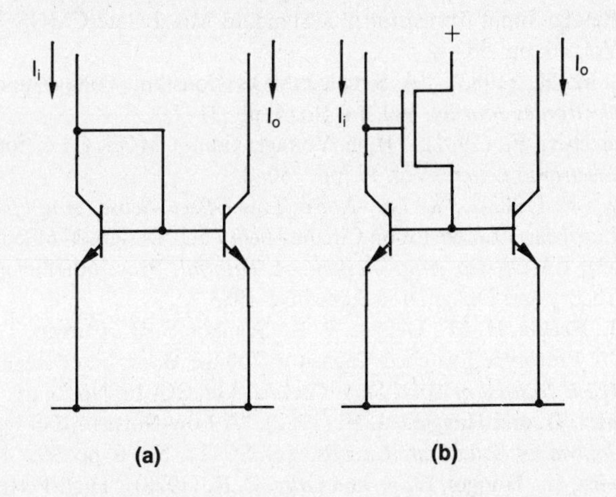

Figure 4.22

4 Using Manufacturer 4 (given in Table 2.2), complete a DC SPICE analysis for the low gain amplifier of Figure 4.19. If gamma (the back gate voltage effect) was to vary from minimum to maximum value, compute the change in output voltage, assuming that both inputs are connected to 2.5 volts.

5 Complete the amplifier design commenced in section 4.4. (Open loop SPICE analysis gives a voltage gain = 1700, an offset voltage = 586 mV, due to output transistors dimensions being in steps of 2.5 μm, current drain = 106 μA, output resistance = 188 kilo-ohms and a dynamic range = −3.5 to 1.8 volts.)

6. A1 pF capacitor is required to stabilize the low gain amplifier shown in Figure 4.19. Do a trial layout of the amplifier. Then, using SPICE and parameters given in Table 2.2, calculate the gain and phase margins.

7 For the amplifier layout undertaken in Question 6, do a transient analysis of the amplifier for the following, using SPICE parameters from Table 2.2:
(a) no load curve;
(b) 5 pF load;
(c) 2 pF in parallel 100 kilo-ohm load.
Assume the input is a step from 2.5 to 2.51 volts.

4.7 References

Fry, P. W. (1969), "A MOST Integrated Differential Amplifier", *IEEE Journal of Solid State Circuits*, Vol. SC–4, pp. 166–8.

Graeme, J. G., Tobey, G. E and Huelsman, L. P. (editors), (1971), *Operational Amplifiers—Design and Application*, McGraw-Hill, New York.

Gray, P. R. and Meyer, R. G. (1982), "MOS Operational Amplifier Design—A Tutorial Overview", *IEEE Journal of Solid State Circuits*, Vol. SC–17, No. 6, pp. 969–82.

Gregorian, R., Nicholson, W. E. (Jr) (1979), "CMOS Switched Capacitor Filters for a PCM Voice Codec", *IEEE Journal of Solid State Circuits*, Vol. SC–14, pp. 970–80.

Gustafsson, S., Sundblad, R. and Svensson, C. (1984), "Low Noise Operational Amplifiers Using Bipolar Input Transistor in a Standard Metal Gate CMOS Process", *Electronic Letters*, Vol. 20, pp. 563–4.

Haskard, M. R. (1983), "A Simple nMOS Constant Voltage and Current Source", *Microelectronics Journal*, Vol. 14, No. 4, pp. 31–7.

Krummenacher, F. (1981), "High Voltage Gain CMOS OTA for Micropower SC Filters", *Electronic Letters*, Vol. 17, pp. 160–2.

McCreary, J. (1983), *"A Low-Noise Low-Offset Sense Amplifier—A Tutorial*, Proc. 6th European Conference on Circuit Theory and Design, 4–6 September 1983.

— (1983), *CMOS Op Amp Design—A Tutorial*, Proc. 6th European Conference on Circuit Theory and Design, 4–6 September 1983.

Mavor, J., Reekie, H. M., Denyer, P. B., Scanlan, S. O., Curran, T. M. and Farrag, A. (1981), "A Prototype Switched Capacitor Voltage Wave Filter Realized in nMOS Technology", *IEEE Journal of Solid State Circuits*, Vol. SC–16, No. 6, pp. 716–23.

Senderowicz, D. and Huggins, J. H. (1982), "A Low-Noise nMOS Operational Amplifier", *IEEE Journal of Solid State Circuits*, Vol. SC–17, No. 6, pp. 999–1008.

Senderowicz, D., Hodges, D. A. and Gray, P. R. (1978), "High-Performance nMOS Operational Amplifier", *IEEE Journal of Solid State Circuits*, Vol. SC–13, No. 6, pp. 760–5.

Smarandoiu, G., Hodges, D. A., Gray, P. R. and Landsburg, G. F. (1976), "CMOS Pulse Code Modulation Voice Codec", *IEEE Journal of Solid State Circuits*, Vol. SC–13, pp. 504–10.

Stone, D. C., Schroeder, J. E., Kaplan, R. H. and Smith, R. A. (1984), "Analogue CMOS Building Blocks for Custom and Semicustom Applications", *IEEE Journal of Solid State Circuits*, Vol. SC–19, pp. 55–61.

Toy, E. (1979), *An nMOS Operational Amplifier*, IEEE International Solid State Circuit Conference, 15 February 1979, pp. 134–5.

Tsividis, Y. (1978), "Design Considerations in Single Channel MOS Analogue Integrated Circuits—A Tutorial", *IEEE Journal of Solid State Circuits*, Vol. SC–13, No. 3, pp. 383–91.

— (1977), "Technique for Increasing the Gain Bandwidth Product of nMOS and pMOS Integrated Inverters", *Electronic Letters*, Vol. 13, No. 4, pp. 421–2.

Tsividis, Y. P. and Gray, P.R. (1976), "An Integrated nMOS Operational Amplifier with Internal Compensation", *IEEE Journal of Solid State Circuits*, Vol. SC–11, No. 6, pp. 748–53.

Wu, B. and Mavor, J. (1983), "High Slew Rate CMOS Operational Amplifier Employing Internal Transistor Compensation", *Microelectronics Journal*, Vol. 14, pp. 5–13.

Young, I. A. (1979), "A High-Performance All-Enhancement nMOS Operational Amplifier", *IEEE Journal of Solid State Circuits*, Vol. SC–14, No. 6, pp. 1070–77.

5 Pseudoanalog techniques

5.1 Important criteria

At this point it would be useful to redefine what is meant by the term "analog circuits". It is assumed that the majority of the on chip processing (filtering, calculating, muliplexing, etc.) will be undertaken in the digital form, involving minimal analog circuitry.

It is essential that any approach to design analog circuits used should:

1. allow the circuits to be accommodated on the same chip as the digital circuit, using the same types of components (transistors, capacitors, etc.) and, preferably, a common supply rail;
2. employ the same standard interface to fabrication that is used for the digital circuits;
3. be insensitive to process parameter variations, particularly when the nMOS fabrication processes are used for multiproject chip work where transistor parameters that are critical for analog circuit operation can have large spreads, (e.g. the depletion transistor threshold voltage and the back gate voltage parameter gamma);
4. use techniques similar to those employed for designing the digital circuits, making it easier for nonelectronic people to include analog circuits, as part of their total system, on a chip.

Pseudoanalog techniques (Haskard and May, 1984) meet all these criteria, for standard inverter circuits are employed with a two phase nonoverlapping clock. The difference is that the inverter circuits are biased in neither the "on" nor "off" state, but in a region of linear operation in between. We will now consider in some detail this approach to analog circuit design for, while not restricted to it, it is a particularly useful method for nMOS work.

5.2 The self biased inverter amplifier

Consider the standard inverter shown in Figure 5.1. When we plot the transfer characteristic (i.e. input versus output voltage), there are three regions, namely, ON, OFF, and the portion in between called the *linear region*. The slope of the linear region is determined in part by the width to length ratio of the two transistors. For an inverter that is OFF, the circuit may be biased to a point such as 1 (shown in Figure 5.1). The input voltage is small while the output voltage is near, if not equal to, V_{dd}. For the ON condition, the bias point may be 2 where the input voltage is large and the output voltage is small in value. In a similar way, if we bias the enhancement transistor by applying a DC input voltage V_i so that the output voltage is V_0, then the inverter is biased into the linear region of operation (bias point 3).

We can illustrate that such a circuit will operate as an amplifier in the following way. If we superimpose on the bias signal V_i a small sinusoidal signal (as shown in Figure 5.2), we can take each instantaneous input voltage (bias plus sinusoidal voltage at times t_1, t_2, t_3, etc.) and plot the output voltage. For small sinusoidal inputs, the tranfer characteristics are approximately linear so that the output voltage is also sinusoidal, but offset from zero by the output bias voltage V_0.

The gain of the inverter stage is given by the slope of the transfer characteristic and is therefore dependent on the dimensions of M1 and M2. Notice

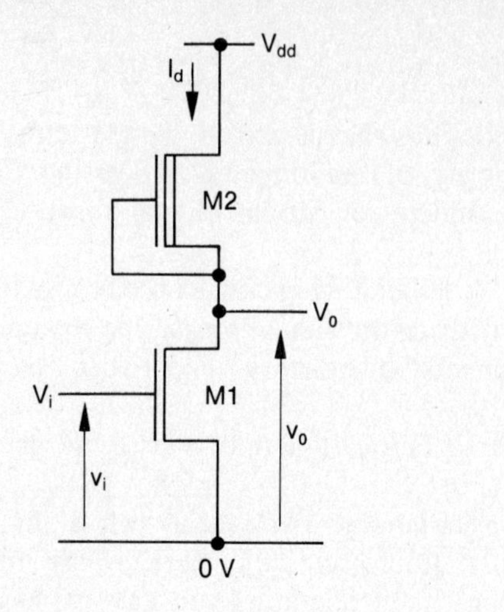

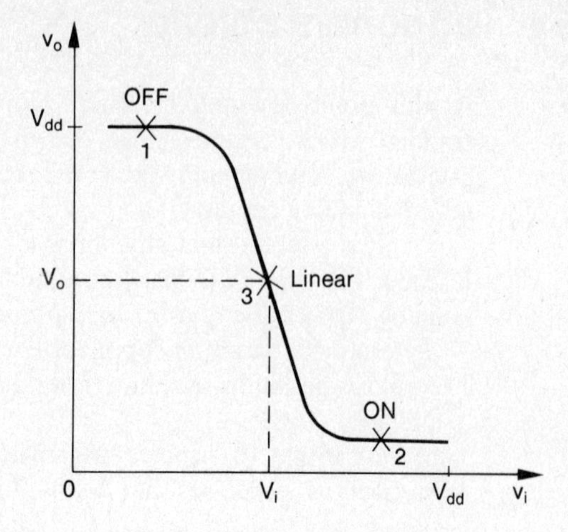

Figure 5.1 Standard inverter circuit

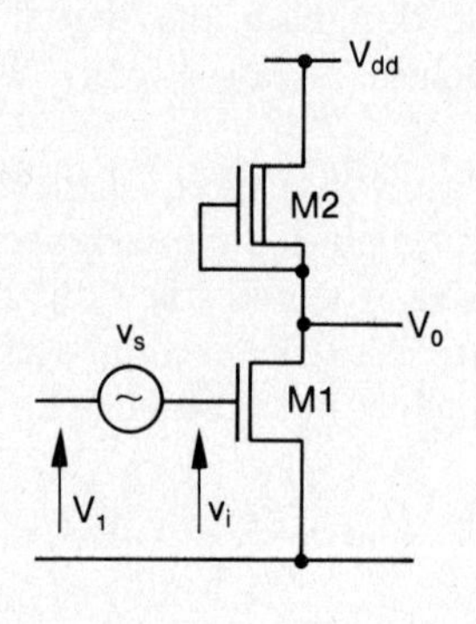

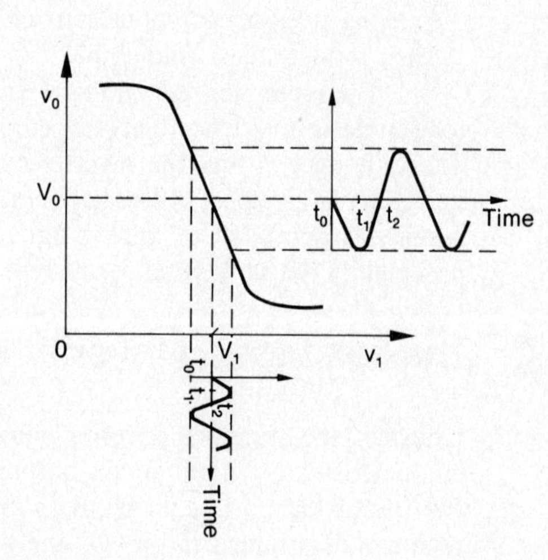

Figure 5.2 Basic inverter as a linear amplifier

that a phase inversion occurs, so that the output sine wave is 180° out of phase with the input.

The actual bias point and voltage gain can be calculated using our simple ideal MOS transistor models. The enhancement transistor M1 will be operating in the saturation region since the drain source voltage under self bias conditions equals the gate source voltage. Thus, using equation 2.13, we obtain the following:

$$I_d = \frac{K}{2}(V_{gs} - V_t)^2$$

$$= \frac{K_1}{2}(V_g - V_{t1})^2$$

The depletion transistor can be operating in either the saturation or nonsaturation region. For the purpose of this calculation, we will assume it is in the nonsaturation region. From equation 2.12 we obtain the following:

$$\begin{aligned} I_d &= KV_{ds}((V_{gs} - V_t) - V_{ds}/2) \\ &= K_2(V_{dd} - V_0)((-V_{t2}) - (V_{dd} - V_0)/2) \end{aligned}$$

Since the drain current is common to the two transistors, we can equate these equations thus:

$$\frac{K_1}{2}(V_g - V_{t1})^2 = K_2(V_{dd} - V_0) \times ((-V_{t2}) - (V_{dd} - V_0)/2)$$

Because this is a quadratic equation, there are two solutions:

$$V_0 = V_{dd} + V_{t2} \pm (V_{t2}^2 - (V_g - V_{t1})^2 K_1/K_2) \qquad \textbf{(5.1)}$$

One of these solutions will not be physically possible, so there is only a single solution. For example, if we take the following practical figures:

$$\begin{aligned} V_{dd} &= 5 \text{ volts} \\ V_{t1} &= 0.7 \text{ volts} \\ V_{t2} &= -4 \text{ volts} \\ K_1/K_2 &= 10 \text{ (ratio of aspect ratios)} \end{aligned}$$

and for self bias,

$$V_g = V_0$$

Then,

$$V_0 = 5 - 4 \pm (16 - 10(V_0 - 0.7)^2)$$

or,

$$11V_0^2 - 16V_0 - 10.1 = 0$$

and,

$$V_0 = \frac{16 \pm \sqrt{(16^2 + 4.11.10.1)}}{22}$$

Since V_0 must be a positive voltage, the following is the only possible solution:

$$V_0 = 1.93 \text{ volts}$$

We can also check that our initial assumption is correct, that is, that the depletion transistor is in the nonsaturation region. We therefore require:

$$V_{ds} < \text{or} = V_{gs} - V_t$$
$$= 0 - (-4)$$
$$= 4$$

Now $V_{ds} = 5 - 1.93 = 3.07$ volts, which is less than 4 volts. Our original assumption was correct.

The gain of the stage can also be calculated from our model by differentiating equation 5.1, the voltage gain of the inverter stage being,

$$A_V = \frac{dV_0}{dV_g} = \frac{-\beta_1/\beta_2(V_g - V_{t1})}{(V^2_{t2} - \beta_1/\beta_2(V_g - V_{t1})^2)^{\frac{1}{2}}}$$

If we insert the typical values used previously, we will obtain the following:

$$A_V = \frac{-10(1.93 - 0.7)}{(16 - 10(1.93 - 0.7)^2)^{\frac{1}{2}}}$$
$$= -13.18$$

The stage gain is negative, indicating a phase change. Further, the gain magnitude of 13.18 is in this case of the same order as the ratio of aspect ratio which was 10. With other process parameters, the gain could be as low as 5 or as high as 20. In other words, the possible spread in gain of an inverter stage may be as high as 4:1, the value depending on the particular process selected for fabrication.

The question that has yet to be answered is how to bias the inverter so that it will always be in the linear region and be independent of process and geometry variations. A simple way of doing this is to bias the transistor, making $V_i = V_0$ (for $v_i = 0$) by momentarily shorting the gate to drain of M1. This process is called self bias (see Figure 5.3). A two phase nonoverlapping clock is employed. During phase 1, transistor M5 is turned on so that $V_{gs} = V_{ds}(= V_i = V_0)$.

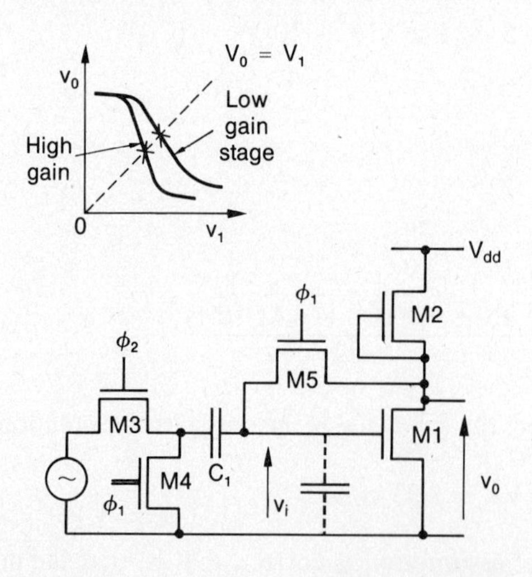

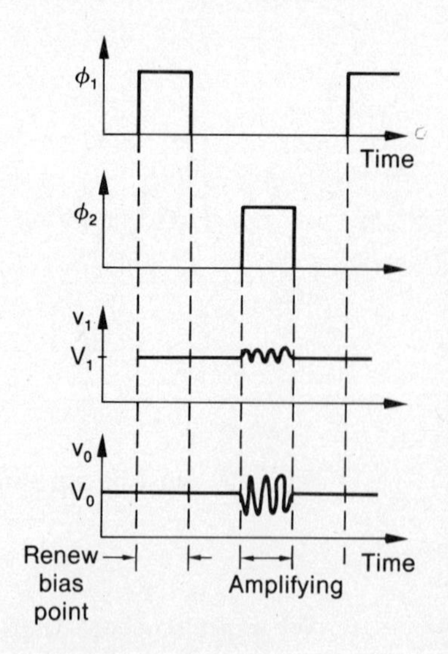

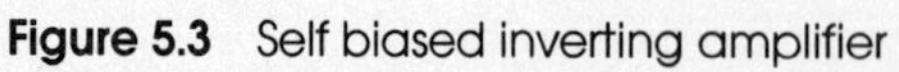
Figure 5.3 Self biased inverting amplifier

Notice that no matter how the inverter transfer characteristic may vary due to process and drawn geometry changes, the inverter is always biased to the linear region. When phase 1 of the clock is removed, the gate capacity holds the gate voltage so that the inverter remains biased to the same point.

The input signal to be amplified is now applied during phase 2 of the clock. The bias isolation capacitor C_i is precharged during phase 1 by switching on M4 so that it will not upset the bias condition. During phase 2, transistor M3 is turned on and the sine wave to be amplified is applied to the inverter amplifier.

Some interesting properties of this circuit will now be explored. During phase 1 of the operation transistor, M4 was switched on to set the voltage across C_1 so that it would not upset the bias condition. We could have connected the source of M4 to a DC potential other than zero and thus have provided an offset voltage. Alternatively, if the clock frequency is much greater than the frequency of signals to be amplified, a second signal can be fed in via M4 to form an algebraic difference amplifier or even difference averaging (Yee et al., 1978), as illustrated in Figure 5.4. Note that for the case (a) of Figure 5.4, when v_i is made zero, v_o is an amplified v_j signal of the same phase.

The basic single input amplifier ($v_j = 0$) provides a 180° phase shift. To increase the gain or restore the phase to zero, two stages of gain can be cascaded. If the geometry of the second inverter is identical to the first, it can be DC coupled to the first stage, since the self bias voltage for both inverters will be the same. This will reduce the silicon area and possibly some of the clock routing problems. Such an amplifier is shown in Figure 5.5.

There are occasions when having the output signal biased to some DC potential (V_O) other than zero is unacceptable. By employing switched transistors on the output, it is possible to DC restore the output to zero volts, or in fact almost any desired potential. Consider the circuit shown in Figure 5.6. During phase 1, when the bias is set up, transistor M8 is also turned on so that the output is set to zero. When the signal is amplified during phase 2, transistor M9 is on so that the signal passes to the output, but is now DC restored to be about zero volts. If the source of

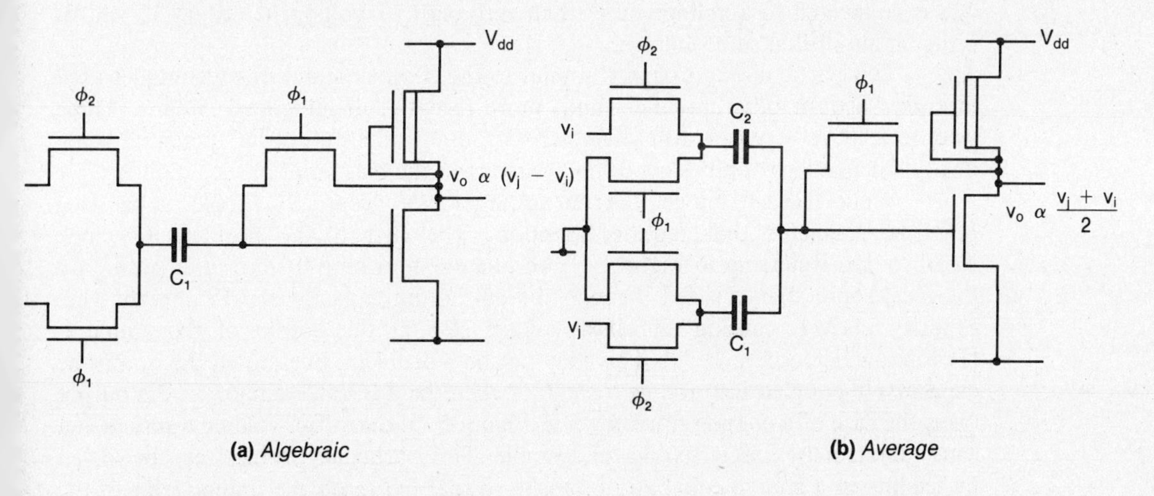

Figure 5.4 Difference amplifiers

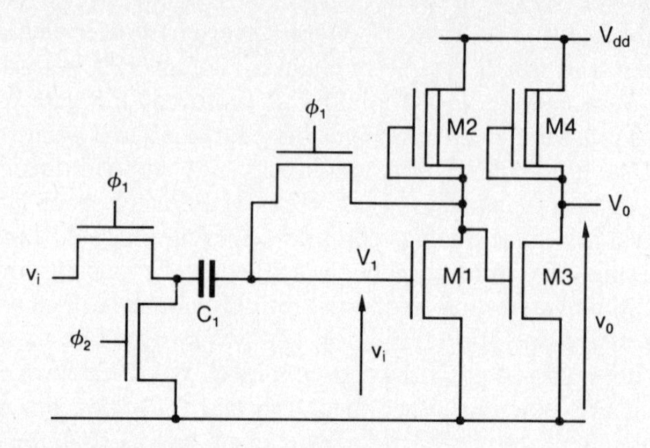

Figure 5.5 Cascaded stages for two identical inverters

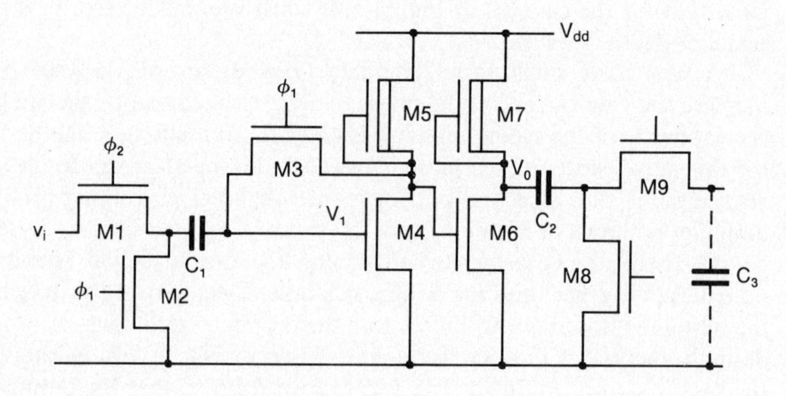

Figure 5.6 Amplifier with DC restoration on the output

M8 is connected to a voltage other than zero (say +1 volts), the output V_o will be centered about that same voltage.

This particular approach is similar to the chopper stabilizing techniques used in early vacuum tube amplifiers and, more recently, in integrated circuits. These were to achieve very low drift characteristics, due to the output DC potential being nearly independent of any slow drifts occurring in the bias point of the amplifier.

While the self biased inverter amplifier is conceptually simple, it has two practical problems that require attention. The first is the problem of clock feedthrough. Referring to Figure 5.3, on phase 1 returning to zero after setting up the bias, some charge is fed through the gate drain capacity of M5 into the gate capacity of M1, causing an offset voltage step at the output of the amplifier (Fotouhi and Hodges, 1979). This may not be a problem, but should the output be capacitively coupled into the next stage or there be a bistable circuit on the output (as is the case of a comparator shown in Figure 5.7), this offset voltage transient can cause an error by falsely setting the bistable. This particular problem can be solved by feeding in a charge equal and opposite to that fed in on the trailing edge of the phase 1 clock, through M3 of Figure 5.7. An extra transistor, M4, is added and

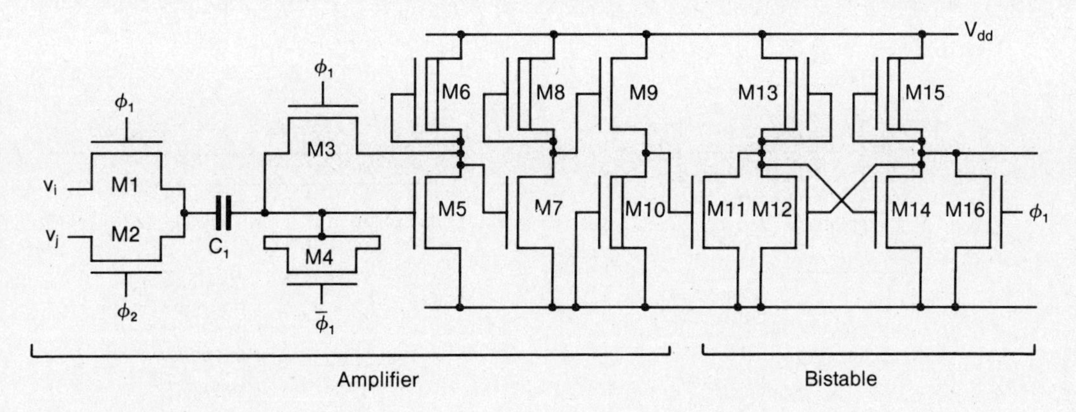

Figure 5.7 Comparator circuit

switched by the complement of phase 1. Since both source and drain are connected to the gate of M5, the gate drain and gate source capacities are in parallel. Consequently, the length of M4 should be the same as M3 but the width must be only half as wide, in order to equalize capacities. Since the gate source drain capacities are extremely small, special attention must be given to the layout of the biasing circuit, otherwise spurious parasitic coupling capacitors will upset the charge balance and cause errors.

Some comment on the operation of the comparator is appropriate here. The amplifier portion is connected as a difference amplifier so that the output is proportional to $(v_i - v_j)$.

During phase 1, when the amplifier bias is set up the bistable circuit (M12–M15) is reset, using M16. During phase 2, if v_i exceeds v_j by a threshold amount (usually designed to be a few millivolts), this signal, when amplified, is able to set the bistable by using the pull down transistor M11. The function of the source follower stage M9–M10 (see section 5.3) is to shift the level of the set up bias signal of the amplifier downward, ensuring that it is insufficient to set the bistable during phase 1.

The second problem with the comparator occurs because series capacitors are employed and every series capacitor has associated with it a parasitic capacitor to substrate (common). This is illustrated in Figure 5.8. The problem is how to connect in the series capacitor C_1 of Figure 5.7 so that the parasitic has minimum effect on circuit operation. Consider the input capacity of the amplifier of Figure 5.9(a). If the parasitic capacity is at the gate end of M4, it will act as an attenuator. Since the parasitic can be comparable in value to the desired capacity, it will introduce a gain reduction of 2. By placing the parasitic at the input (MUX or M1/M2) end this problem will be solved. If, however, there is a capacitor on the output of the amplifier (as shown in Figure 5.9(b)), the parasitic should be at the amplifier output end, otherwise a similar gain reduction will occur.

If either a depletion transistor gate capacity or poly to diffusion capacity is used as the series capacitor, the parasitic capacities will be diode capacities. In that case care must be taken that any signal transient step does not force the diode into forward conduction, as this will upset normal circuit operation.

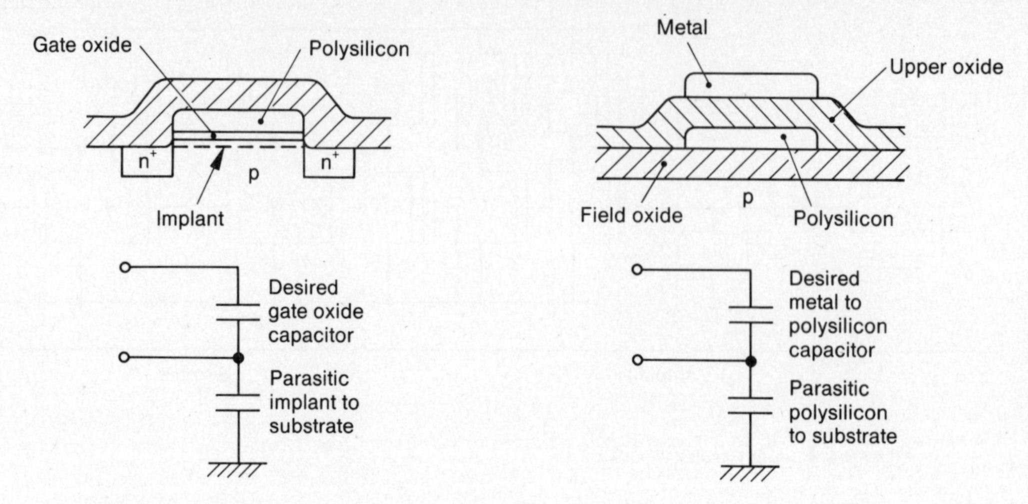

Figure 5.8 Floating capacities in the nMOS process have a large parasitic capacity associated with them

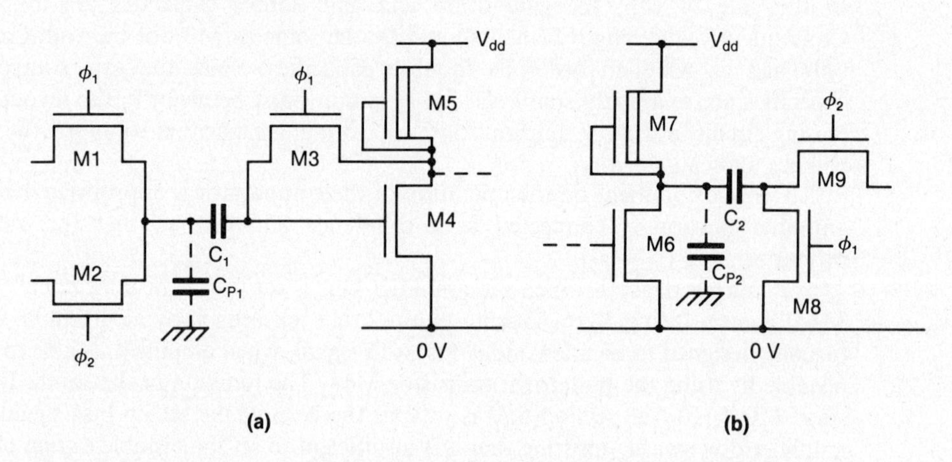

Figure 5.9 Orientation of series capacitor so parasitic has minimal effect

5.3 Alternative circuits

In addition to the standard self biased inverter circuit, there are three other configurations that are useful for pseudoanalog techniques (Haskard, 1983). The first two are shown in Figure 5.10, with part (b) of the figure showing two variations of one circuit.

The standard inverter previously discussed has the advantage in that it offers the maximum gain of all the possible combinations of two transistors. When low voltage gains or constant voltage sources are required, the circuit given in Figure 5.10(a) is important. By connecting the depletion transistor gate to the common rail, negative feedback is applied, which stabilizes both the gain and bias voltages. As an

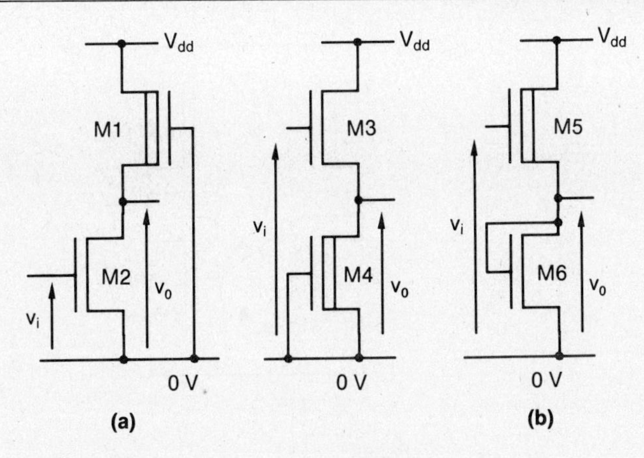

Figure 5.10 Useful circuits for pseudoanalog techniques

example, Table 5.1 provides SPICE simulation figures for a unity gain inverting amplifier, for two very different nMOS processes. Although bias voltages and current drains differ significantly, the voltage gains are both unity.

Table 5.1 Unity gain inverting amplifier as per Figure 5.9(a)

Manufacturer	*Current drain* (μA)	*Output* DC *bias voltage* (V)	*Voltage gain*	*Output resistance* (kilo-ohm)
1	23.8	1.68	−1.01	22.2
2	39	2.09	−1.01	18.1

Notes:
1. Ratio of aspect ratios is 1.75.
2. SPICE information is from Table 2.2.

The second configuration (Figure 5.10(b)), which is known as a source follower, has already been dealt with as part of the comparator circuit. It provides a less than unity voltage gain, a low output impedence for driving large capacity loads, and level shifts any input voltage by the upper transistor gate source voltage toward zero volts. Where low impedance is required, with minimum level shift, the transistor types can be reversed (as shown in the alternative circuit).

The final circuit of interest is a modified bistable (shown in Figure 5.11). If all transistors are turned off through M5, then when M5 is turned on, the loop gain of the bistable is such that only a small signal need be fed into either of the gates (v_i and/or v_j) to bias the circuit to a preferred state. Such a circuit could be used as a comparator or squaring circuit.

5.4 Using a tile approach

If one examines the circuit configuration discussed in this section and Chapters 6, 7 and 8, there are many similarities. In fact, in many instances, the same basic

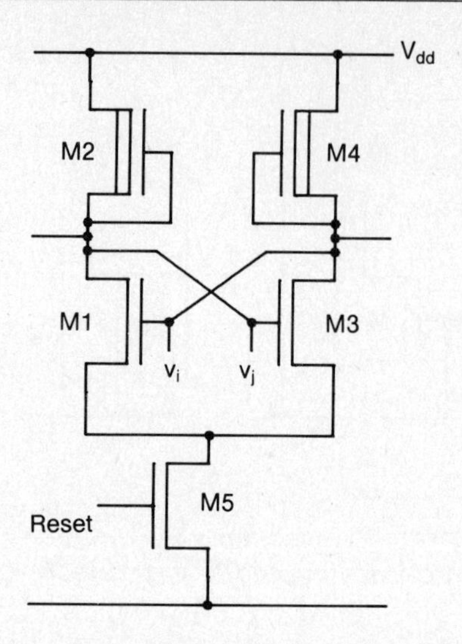

Figure 5.11 High gain pseudoanalog circuit using regeneration

components are employed, but they are simply rearranged (Figure 5.12). To simplify the design of these circuits, a set of primitive cells (called *tiles*) has been designed (see Appendix A). These can be assembled to form a desired circuit called a *mosaic* (Haskard and May, 1985). Each primitive cell is of a standard size of 12

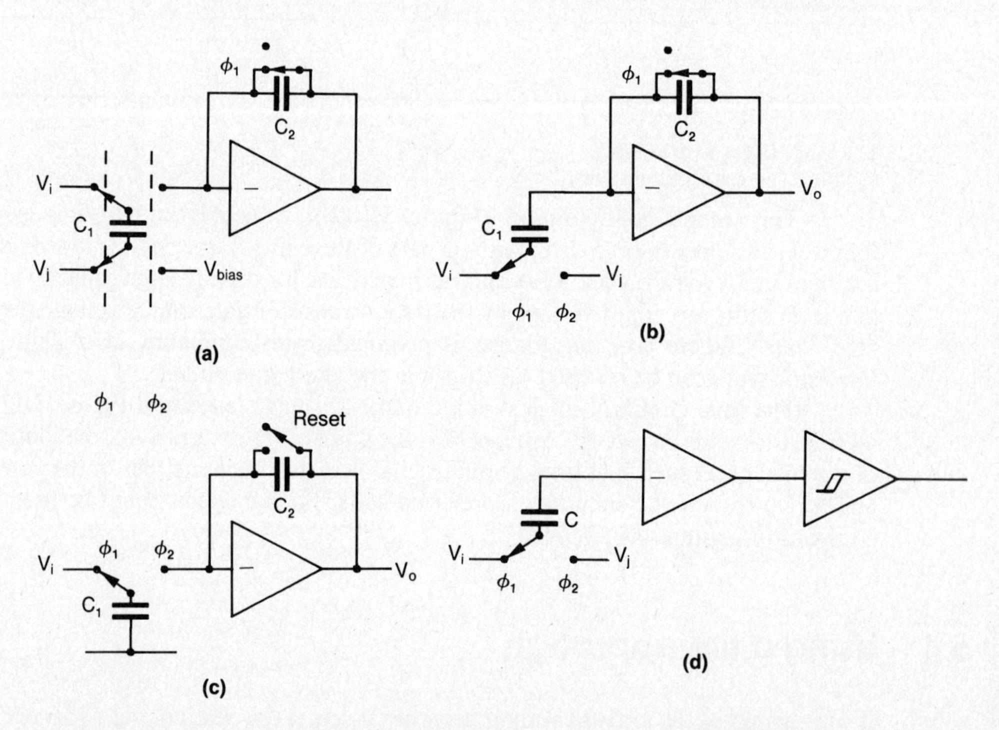

Figure 5.12 Typical analog circuits

lambda by 28 lambda or, in a few cases, a multiple of one of these standard dimensions. The more complex mosaics are formed by laying the tiles side by side, interconnections occurring automatically through the abutment process. To give the system flexibility, each tile has at least one stretch position in both the X and Y directions so that tiles can be expanded to accommodate any grid size larger than the minimum size.

A necessary part of tile/mosaic generation is the defining of signal layers. Signals flowing horizontally are metal and those flowing vertically are polysilicon. Consequently, power rails run horizontally and clocks run vertically, while signal tracks can be metal or polysilicon, depending on the direction. Clock lines are centrally located to allow mirroring of tiles.

It is interesting to note that since the basic amplifier tile is a standard inverter (4:1, 6:1 or 9:1 ratios being available when calling the cell), the tile concept can be extended to the digital area, with a complementary set of digital tiles forming a mosaic.

Wherever appropriate, any library cell should be parameterized so that the user need simply to specify that parameter as being part of the cell call definition. The gain of an amplifier or the number of bits for a converter are examples of this. Naturally, these parameters have limits (both upper and lower) that cannot be exceeded.

Layout drawings for some tile cells are given in Appendix A, with the inverter amplifier, series and shunt capacitors, series transmission transistors, feedthroughs, bias circuits and a reset bistable cell included. Figure 5.13 illustrates how three tiles may be combined to form a simple self biased single stage amplifier.

Further and more detailed examples of the use of these and other tiles will be given in later chapters of this book.

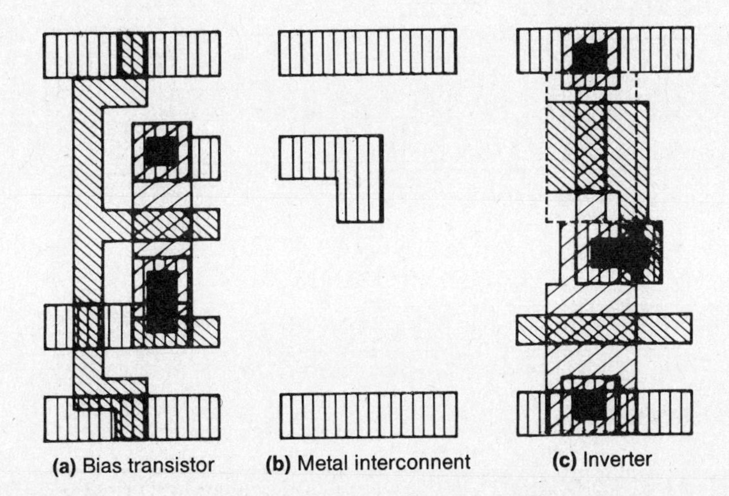

Note:
(a) abuts (c) and (b) overlays (c) to form a self biased inverter

Figure 5.13 Example of how cells may be combined

5.5 Exercises

1 A three stage amplifier, each stage being identical, may be self biased, using the two methods shown in Figure 5.14. Which method would you prefer and why? Could both methods be employed for a two stage amplifier? If not, why not?

2 Prove by simple analysis that the circuit diagrams of Figure 5.4 do perform algebraic and average differences.

3 Care must be taken in laying out a self biased inverter stage, otherwise stray capacities can couple in small charges giving rise to offsets. Assuming the drain/source diffusion extends 0.5 μm under the gate and using the capacity figures in Table 2.5, calculate:
(a) the gate to drain capacity of a minimum geometry enhancement transistor; and
(b) the stray capacity between minimum widths metal and polysilicon lines.
With reference to Figure 5.3, if the enhancement transistor of part (a) is M5 and the stray capacity of part (b) is from the phase 1 metal to M1 polysilicon gate line, estimate the effect of the stray capacity on the biasing of the inverter.

4 Using the tiles given in Appendix A, produce a layout for the comparator circuit given in Figure 5.7.

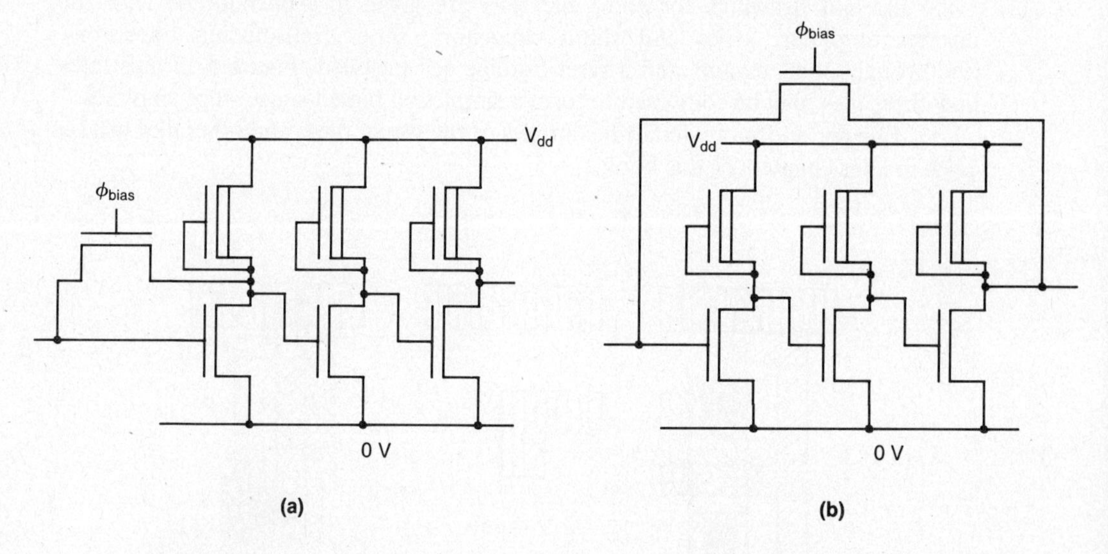

Figure 5.14

5.6 References

Fotouhi, B. and Hodges, D. A. (1979), "High Resolution A/D Conversion in MOS LSI", *IEEE Journal of Solid State Circuits*, Vol. SC–14, pp. 920–6.

Haskard, M. R. (1983), "Experiments Towards a Standard Interface to Fabrication for Analogue Circuits", *Technical Report VLSI-TR-83-1-1*, Commonwealth Scientific and Industrial Research Organisation, Australia, January.

Haskard, M. R. and May, I. C. (1985), "A Library of Analogue Cells for the System Designer", *Journal of Electrical and Electronic Engineering* (Australia), Vol. 5, pp. 262–77.

— (1984), "A Library of Analogue Cells for the System Designer", *VLSI PARC Conference Digest* (Aust.), 15–17 May 1984, pp. 65–6.

Yee, Y. S., Terman, L. M. and Heller, L. G. (1978), "A 1 mV MOS Comparator", *IEEE Journal of Solid State Circuits*, Vol. SC–13, pp. 294–8.

6 Amplifiers and filters

6.1 Charge flow concepts

Discussions in the earlier chapters have narrowed the useful components for analog work to capacitors, transistors employed as switches, inverters biased to provide amplification and, with careful design, unity and higher gain conventional operational amplifiers. While the list may appear restrictive, it is still possible to design the majority, if not all, of the analog amplifier class of circuits required.

The use of charge is concentrated on here, rather than the use of current (Fudim, 1984). Consider the situation when a resistor R is connected across a voltage V. The current that flows is given by the simple relationship of Ohm's Law, according to which,

$$I_R = \frac{V}{R} \tag{6.1}$$

If now a capacitor C is placed across the voltage, the current flowing is time dependent, namely,

$$I_C = C\frac{dV}{dt}$$

However, if charge rather than current is considered, the charge created in the capacitor is,

$$Q = \frac{V}{(1/C)} \tag{6.2}$$

Comparing this to equation 6.1, we see that there is a similar linear relationship. To make use of this charge approach, a continuous flow of current must not occur. Thus the capacitors must be constantly switched while parasitic leaks of capacitors must be minimized, otherwise significant errors will result. Further, the ON resistance of the transistor switch will limit the flow of the charge. As a result, the switch must remain closed allowing sufficient time for the charge to transfer. This restriction will limit the maximum switching rate of the circuit.

Some comment is therefore needed on the choice of transistor switch. In the case of nMOS, there is no choice, with only n-channel enhancement transistors being available. This is not so with CMOS, for at least one other option exists, namely back to back n- and p-channel enhancement transistors. The advantage of this circuit is that the ON resistance is low and near constant over the whole range of the control signal (see Exercise 2 in Chapter 3). However, the disadvantage of this circuit is that a complementary clock line is required to switch the p-channel transistor, requiring that extra inverters or additional clock lines be added to the chip layout. In most cases, this additional complexity is not warranted and the single enhancement n-channel transistor by itself is used as the switch.

The switched capacitor approach allows both charge transfer and charge summing, which leads to a wide range of useful circuits that will be considered in this and later chapters. An example of charge summing is the weighted capacitor digital to analog converter of Figure 7.5 (p. 119).

6.2 Gain controlled amplifiers

With conventional operational amplifiers, the gain can be controlled by passive feedback components, normally the ratio of two resistors (as shown in Figure 6.1). It is assumed that A is much greater than A_v. Similar techniques can be employed using switched capacitor methods, which are a further example of charge summing. Consider the pseudoanalog amplifier circuit given in Figure 6.2. Initially the input capacity C_1 is grounded and capacitor C_2 is short circuited. For an ideal amplifier, the total charge on the two capacitors is C_1V_B and the output voltage from the circuit is V_B. On simultaneously opening switch S_2 and changing over switch S_1 to the input voltage v_i, the total charge on the two capacitors is:

$$C_1(V_B - v_i) + C_2(-v_o)$$

Equating the charges before and after switching gives the following gain:

$$A_v = -\frac{C_1}{C_2} \tag{6.3}$$

This derivation assumes A is very large so that V_B remains essentially constant. If this is not the case, then,

$$A_v = \frac{-C_1}{\frac{C_1}{A} + \left(C_2 \frac{A+1}{A}\right)} \tag{6.4}$$

and an error has been introduced.

In the normal situation, the amplifier is designed so that equation 6.3 holds and the gain is controlled by the ratio of two capacitors. Further, if a conventional operational amplifier has been used, the analysis is identical, with V_B being the input offset voltage.

As has been previously explained in Chapter 5, the inputs to switch S_1 can be reversed so that initially it is connected to the input signal v_i and then on switching grounded. Here the gain is of the same magnitude as given by equation

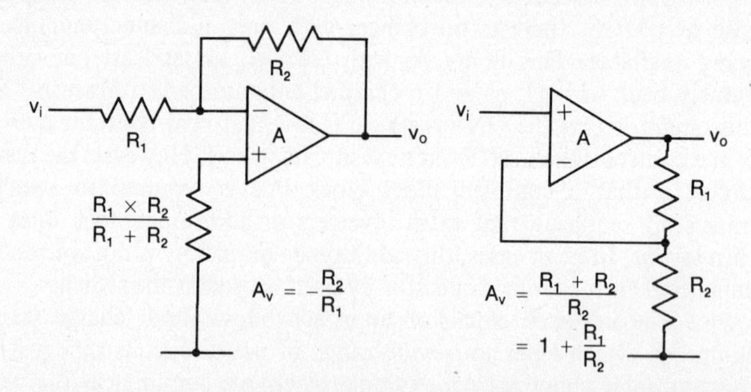

Figure 6.1 Standard gain controlling methods for conventional operational amplifiers

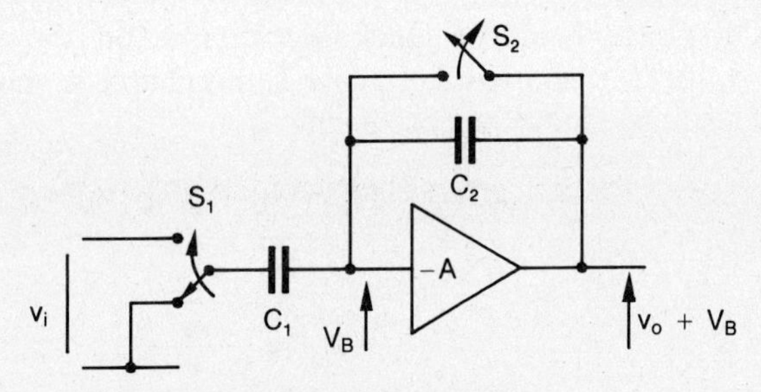

Figure 6.2 A gain controlled amplifier using capacitors

4.3 but there is no phase reversal. Alternatively, a differential mode (Figure 6.3) can be used, giving:

$$A_v = \frac{V_o}{V_j - V_i}$$

$$= \frac{C_1}{C_2}$$

All signals are defined in Figure 6.3(a). A difficulty with this circuit is that if S_1 is driven by a two phase nonoverlapping clock, signals v_j and v_i must be available on different clock phases. An alternative approach is to switch both terminals of capacitor C_1 (as shown in Figure 6.3(b)).

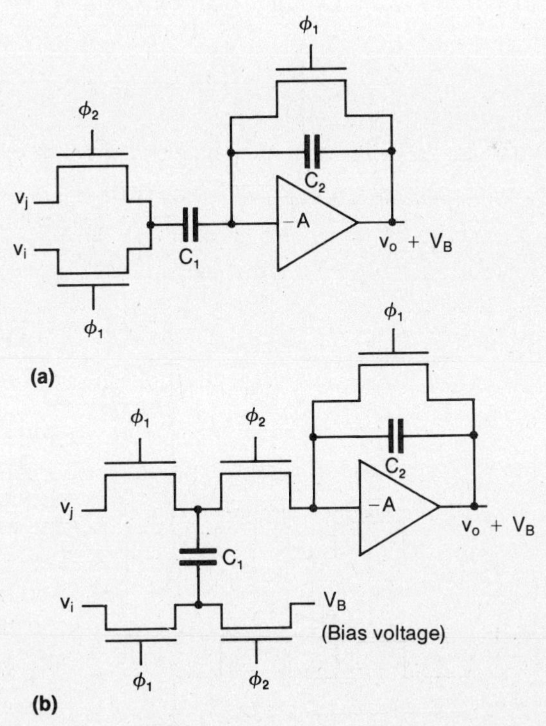

Figure 6.3 Two methods of implementing controlled gain differential amplifiers using pseudoanalog techniques

On phase 1 of the clock, the charge on the two capacitors is simply $C_1(v_j - v_i)$ as C_2 is shorted. On phase 2, this charge is applied to the amplifier input with the resulting charge balance giving,

$$C_1(v_j - v_i) = C_2(-v_o)$$

Thus,

$$A_v = \frac{V_o}{V_j - V_i} = -\frac{C_1}{C_2} \tag{6.5}$$

Note that for a pseudoanalog amplifier on phase 2, the bottom end of C_2 must be connected to a voltage V_B, otherwise an unwanted offset is introduced. Since this voltage is the self bias voltage of an inverter, it is easily generated, using a separate inverter or, if the system employs amplifiers operating on alternative phases, the bias voltage can be derived from an amplifier being self biased on phase 2.

An alternative way to do this for a unity gain amplifier is to use a self compensating method, as illustrated in Figure 6.4. Here, during phase 1, capacitor C_2 is charged to the self bias voltage V_B. The resultant gain expression is,

$$\begin{aligned} A_v &= \frac{V_o}{V_j - V_i} \\ &= -\frac{C_1}{C_2} - \frac{V_B(C_2 - C_2)}{(V_j - V_i)C_1} \\ &= -\frac{C_1}{C_2} \\ &= -1 \text{ for } C_1 = C_2 \end{aligned}$$

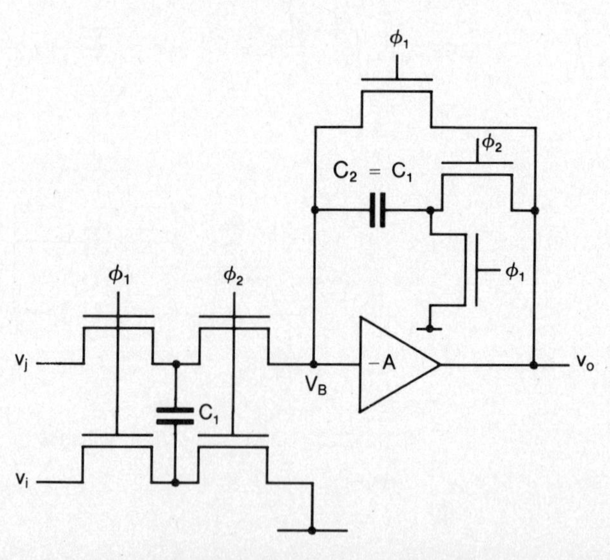

Figure 6.4 Self compensating unity gain switched capacitor amplifier

Where conventional operational amplifiers are employed, the self compensating method can also be used to cancel offset. Additional methods have been reported (see Temes, 1984; Gregorian, 1982). Figure 6.5 shows the circuit of an amplifier employing standard tiles. The open loop gain is typically 2000 while the capacitor ratio controls the gain to 2. Table 6.1 provides more detailed results measured on the amplifier.

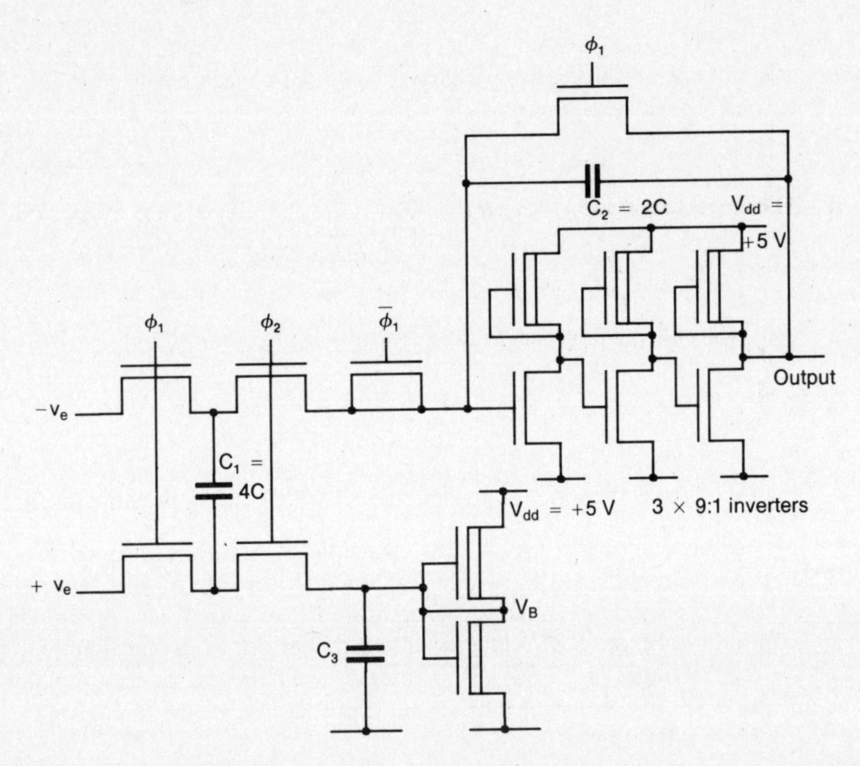

Figure 6.5 Gain controlled switched capacitor amplifier made from the tile library

Table 6.1 Performance figures for the switched capacitor amplifier given in Figure 6.5

Parameter	*Measured result*	*Units*
AC gain	1.95	—
Common mode input range	0 to 3.5	volts
Maximum clock frequency	0.5	MHz
Self bias voltage V_B	1.8	volts
Maximum output swing range	0.2 to 4.8	volts
Small signal output impedance	2.5	kilo-ohm

Notes:
V_{dd} = 5 volts
Clock frequency = 100 kHz
Input signal frequency = 1 kHz
(unless otherwise noted).

6.3 Simulating a resistor

An interesting by product of charge transfer, using a switched capacitor, is that such a circuit can be made to look similar to a resistor. Consider Figure 6.6(a). C is alternatively connected to v_i and v_j. Each time a charge Δq flows where,

$$\Delta q = C(v_i - v_j)$$

due to C being alternatively charged to v_i and v_j each clock period T. The average current that flows is, then,

$$i_{avg} = \frac{\Delta q}{T} = \frac{C(v_i - v_j)}{T}$$

while the equivalent resistor that seems to exist between v_i and v_j is,

$$R_{eq} = \frac{T}{C} \tag{6.6}$$

Figure 6.6(b) is the series equivalent circuit which yields the same result.

We can make use of this principle to form an alternative type of gain controlled amplifier, as shown in Figure 6.7. More importantly, we can employ this simulated resistor to make an integrator or low pass filter, as shown by the two examples in Figure 6.8. While these circuits perform a similar function, they are not the same (Gregorian, 1983). For example, while there is a delay, equal to the period of the clock (T), between output and input for circuit (a) of Figure 6.8, for circuit (b), there is no delay.

The practical implementation of these circuits is not as simple as it may at first seem. The switching transistors are not ideal so clock feedthrough occurs via the gate source/drain capacities. Providing the feedback capacitor C_2 is comparatively large this feedthrough should not be a problem (Caves et al., 1977). Parasitic

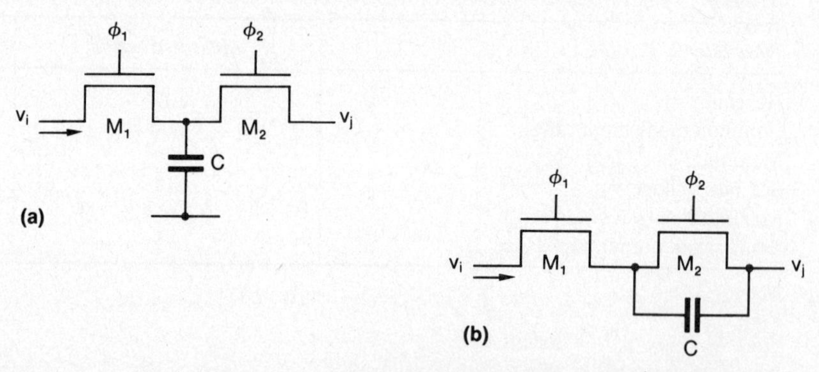

Figure 6.6 Switched capacitor simulated resistor circuits

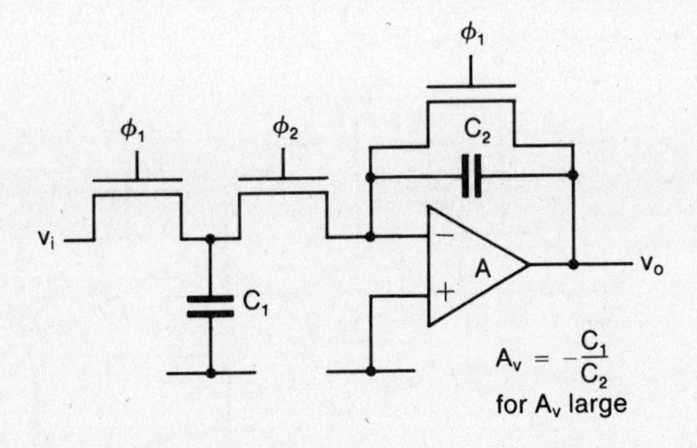

Figure 6.7 A gain controlled amplifier

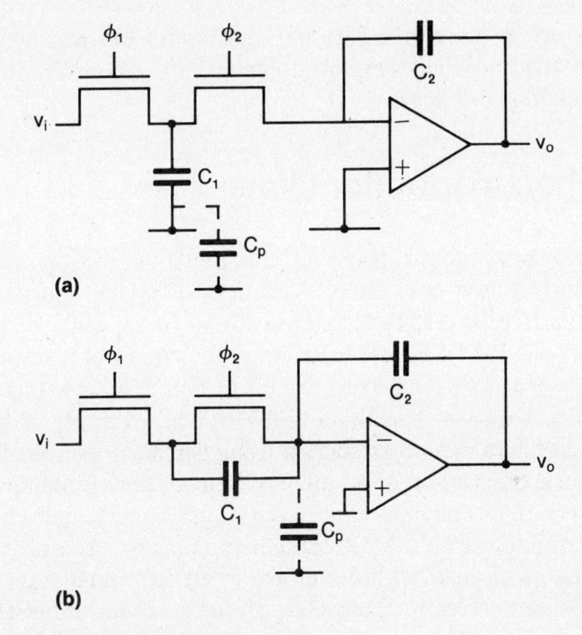

Figure 6.8 Switched capacitor integrators

capacities to substrate associated with the source/drain of the switching transistors can also introduce significant errors. This can be overcome by increasing the desired capacitor area to make the parasitics small in relative value or by employing more complex circuits that are less parasitic prone. Such a circuit (Gregorian, 1983) is shown in Figure 6.9. Analysis has shown that for an infinite gain amplifier, none of the parasitic capacities associated with either C_1 or the switches will contribute an error charge to the feedback capacitor C_2 if both ends of the shunt capacity C_1 of the circuit (Figure 6.8(a)) are switched.

Ghausi (1984); Faruque et al. (1982); Bermudez and Bhattacharyya (1982) and Raut and Bhattacharyya (1984) provide further examples of circuits that are insensitive to stray capacitance.

Before concluding discussion on this source of error, it should be noted that neither of the two circuits given in Figure 6.8, if connected up correctly need be

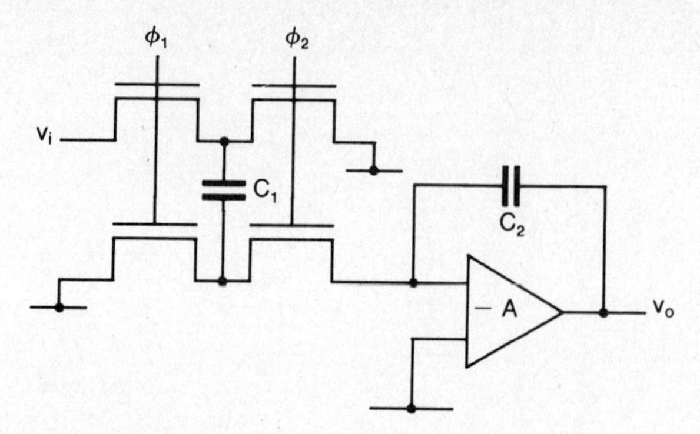

Figure 6.9 Stray or parasitic capacitor insensitive integrator

prone to errors due to the parasitic capacity to substrate which is associated with C_1. Further, the amplifiers employed can be either conventional or, with some special care, pseudoanalog types.

6.4 Switched capacitor filters

The integrator circuits given in the previous section are first order networks, performing as low pass filters. Networks of higher order can be formed by many techniques (see Baker, 1984; Saraswat and Saha, 1984; Fleischer and Laker, 1979; Mavor et al., 1981). Figure 6.10 shows the circuit of a second order network, based on the series switched capacitor circuit of Figure 6.8(b) (Baker, 1984).

See Saraswat and Saha (1984) for an example of how low pass, high pass and notch filters can be assembled using cascaded sections and Fleischer and Laker (1979) for a description of a family of biquad/filter building blocks.

The filter circuits that require only unity gain buffer amplifiers are of particular interest to nMOS designers, mainly because these amplifiers can be fabricated more easily (Mavor et al., 1981; Raut and Bhattacharyya, 1984; Herbst and Hosticka, 1980). Examples of unity gain amplifiers of the conventional operational amplifier type (Chapter 4) and pseudoanalog switched capacitor type have already been given earlier in this chapter. They may be employed in these circuits. Figure 6.11 shows the circuit of the basic element that can be used to assemble voltage wave filters. (See Mavor et al., 1981, for details on nMOS designs up to seventh-order filters.)

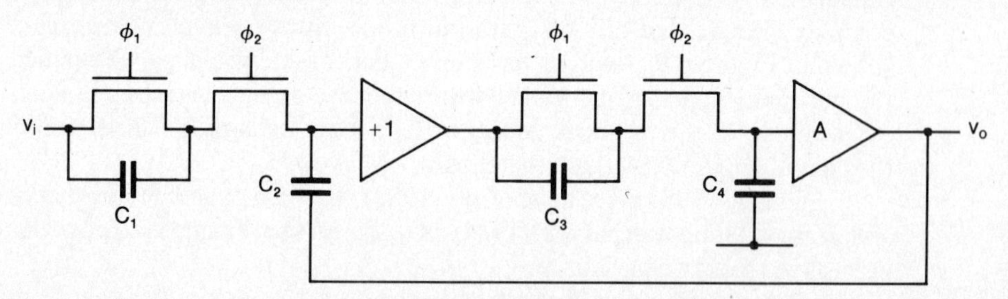

Figure 6.10 A second order switched capacitor filter

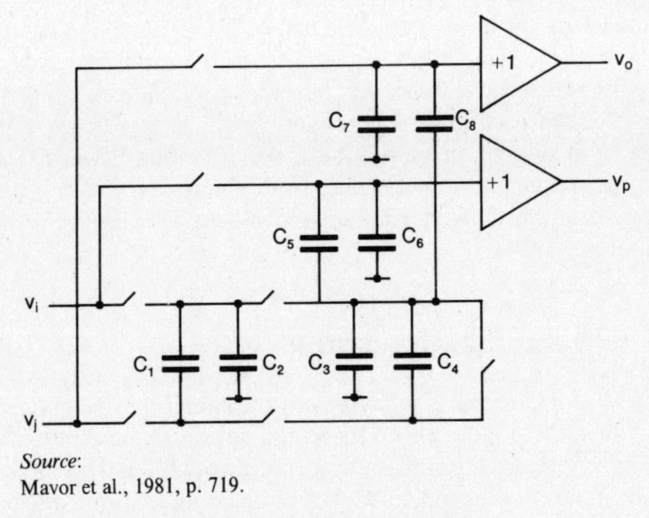

Source:
Mavor et al., 1981, p. 719.

Figure 6.11 Basic building block for a voltage wave filter

Raut and Bhattacharyya (1984) give an example of a band pass filter circuit that is insensitive to parasitics and employs only unity gain buffer amplifiers (Figure 6.12).

Switched capacitor filters do have a number of problems. First, they tend to be noisier than their conventional bipolar analog counterparts (see Gobet and Knob, 1983; Fischer, 1982). Sources of this noise are the thermal noise from the on

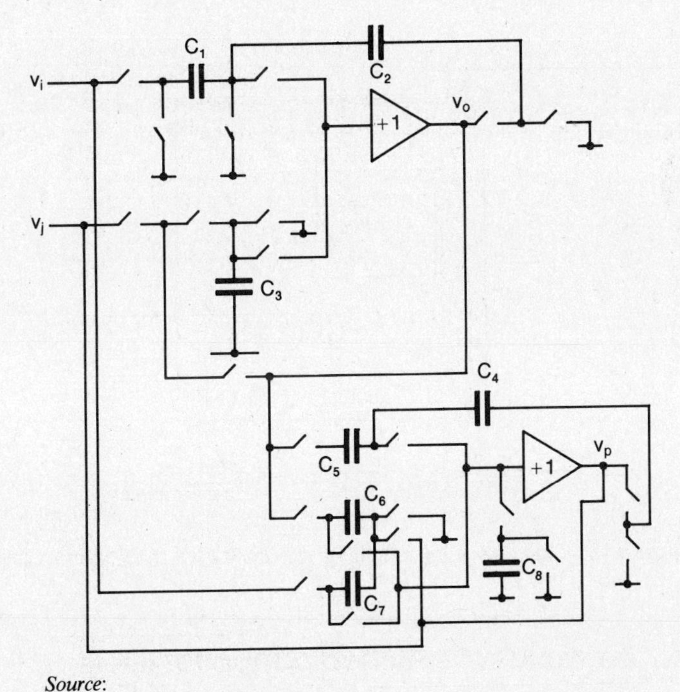

Source:
Raut and Bhattacharyya, 1984, p. 110.

Figure 6.12 A biquad bandpass filter using unity gain buffer amplifiers

resistances of the transistor switches (= kT/C) and the more dominant amplifier noise. While problems arise mainly from the input stage, because stage gains are often low, subsequent stages can also make a significant contribution. Second, there is the problem of amplifier finite bandwidth. In a switched capacitor filter, the pole frequency is directly proportional to the switching frequency. Consequently, as the switching frequency–amplifier bandwidth ratio is increased, the pole frequency and Q vary from those predicted. Active compensation can be used to reduce this effect (Sasikumar, 1983).

6.5 Ground rules for layout

The final problem area is in laying out the circuit on silicon (Gregorian, 1982). Since stray capacities couple all nodes to the substrate, care must be taken to minimize noise injection into the substrate, thence through the strays into the signal path. The bottom plates of all capacitors should be connected or switched only to ground or low impedence points, such as an amplifier's output.

Where analog and digital circuits are to be on the same chip, it is essential to keep power lines separated to prevent coupling. A diffusion of opposite type to the substrate can be run around the analog circuit area and connected to a supply rail to reverse bias the diffusion substrate diode. Separate the circuits as much as possible (Figure 6.13). To neutralize clock feed through to sensitive circuit nodes, a dummy transistor of half dimensions can be employed and fed from the offending clock complement signal.

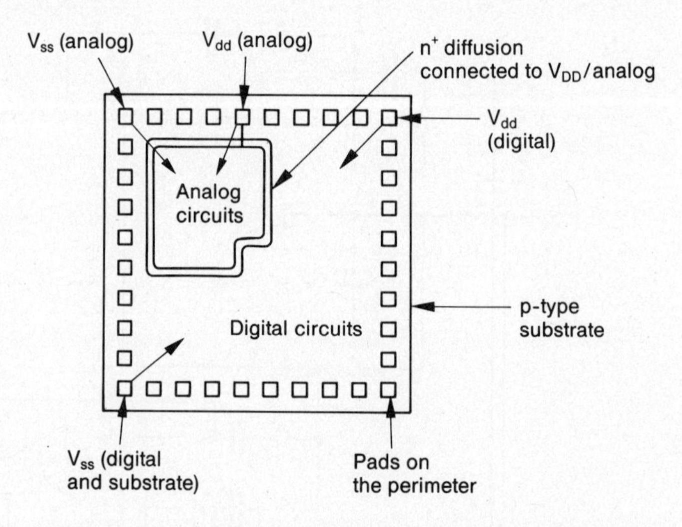

Figure 6.13 Chip layout illustrating separation of analog and digital circuits

6.6 Examples of switched capacitor filters

Figure 6.14 shows the circuit of a simple first order low pass switched capacitor filter which uses the "BUFAMP" buffer amplifier (Figure 4.9) as an output driver. This is

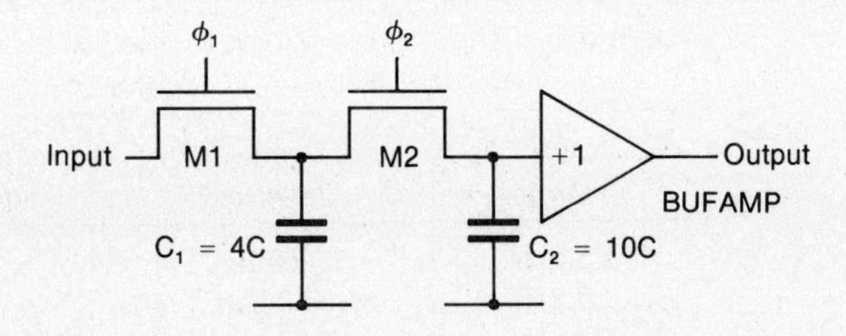

Figure 6.14 A simple first order low pass filter

about the simplest circuit possible and consequently has some serious drawbacks, the biggest problem being poor stop band performance. The parasitic drain source capacities of M1 and M2 couple high frequencies directly through to the amplifier, with clock feedthrough occurring due to the gate source and gate drain capacities. The clock feedthrough increases with the clock frequency, while the signal feedthrough increases with the signal frequency. Table 6.2 gives performance details of the filter.

Table 6.2 Performance figures for the first order low pass filter of Figure 6.14

Clock frequency	*Cut-off (−3 db) frequency*	*Stop band attenuation*
2.5 kHz	140 Hz	12 db
7.5 kHz	410 Hz	12 db
25 kHz	1.3 kHz	12 db
50 kHz	2.7 kHz	12 db

The circuit shown in Figure 6.15 is another first order low pass filter; the capacitor switching, however, has been arranged to minimize the effects of parasitic capacities. This filter has twice (in db) the stop band attenuation of the circuit of Figure 6.14. Table 6.3 gives the performance details.

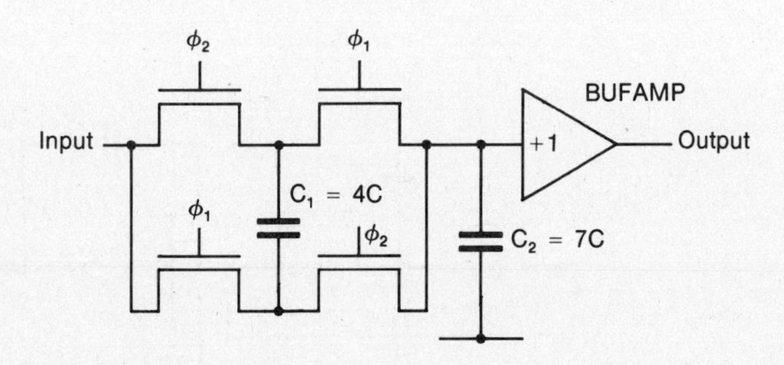

Figure 6.15 An improved first order low pass filter

Table 6.3 Performance figures for the first order low pass filter circuit in Figure 6.15

Clock frequency	*Cut-off (−3 db) frequency*	*Stop band attenuation*
2.5 kHz	460 Hz	27 db
7.5 kHz	1.3 kHz	25 db
25 kHz	4.3 kHz	18 db
50 kHz	8.5 kHz	16 db

6.7 Additional switched capacitor circuits

Switched capacitor methods can be used to perform functions other than filtering, conversion and amplification. Combinations of these functions are, of course, possible. Figure 6.16 shows a standard summing integrating circuit that is frequently employed as a general purpose analog building block.

The concepts can be extended (Fudim, 1984; Yasumoto and Enomoto, 1982; McCharles and Hodges, 1978) to include circuits that perform the functions of sample and hold, four quadrant analog multiplication, pulse to voltage conversion, digitally controlled resistance, and frequency multiplication and division. We will not discuss these operations, as they can be performed digitally.

The range of circuits can be further increased if simple analog delay lines are included (Tsividis, 1980; Howes and Morgan, 1979). These can be constructed using capacity bucket brigades or cascaded unity gain amplifiers. While the charge transfer efficiency of the capacity bucket brigades is poor so lines cannot be long, cascaded amplifiers occupy a large silicon area. A combination of the two can in some instances be a satisfactory compromise where amplifiers are periodically inserted into the bucket brigade to restore the signal level. Figure 6.17 shows a simple bucket brigade circuit using a three phase clock. On phase 1, the transfer capacitor C_1 is discharged. On phase 2, the voltage is clocked from the preceding stage into the transfer capacitor, followed by phase 3, where it is passed into the storage capacitor C_2. Thus, for each clock cycle the signal shifts one stage. Delays of integer multiples of the clock period are achieved. The source followers and switches

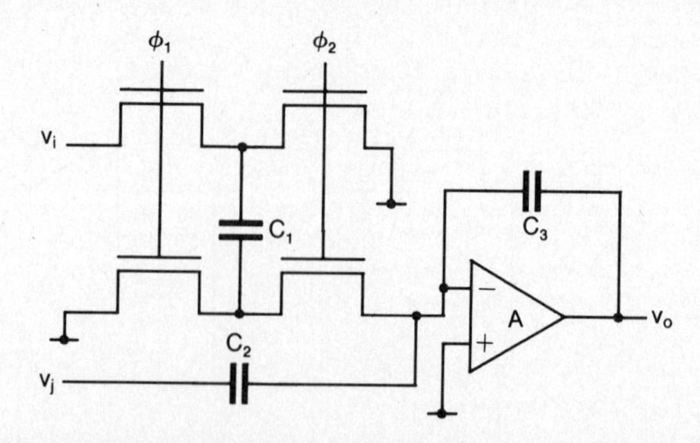

Figure 6.16 General purpose analog building block

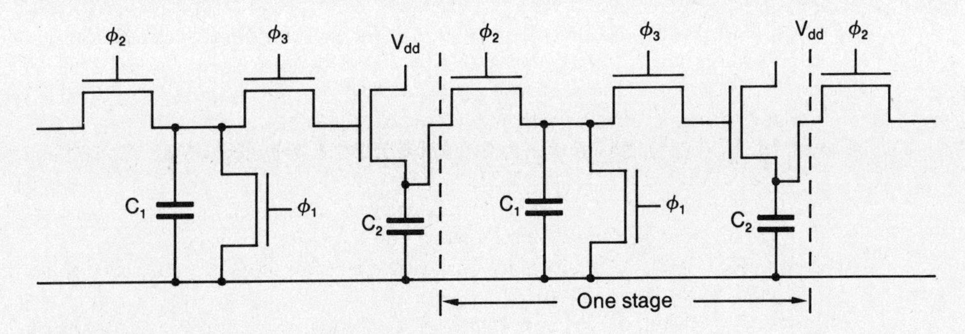

Figure 6.17 A bucket brigade delay line

all attenuate the signal by at least V_t. The source followers can be omitted, but the signal is still attenuated through charge sharing.

6.8 Exercises

1 Refer to Figure 6.2.
   (a) For the case when the amplifier gain A is not infinite, derive an expression for the circuit gain (equation 6.4).
   (b) If the input capacity to the amplifier C_A is included in the circuit and if A is finite, calculate the error that this input capacity C_A introduces.

2 Refer to Figure 6.3(b). If the bias voltage is made zero, derive an expression for the input output voltage from the circuit.

3 Refer to Figure 6.8(b). Assuming a clock period of T, derive an expression for v_o the output voltage.

4 Repeat question 3 but for the circuit given in Figure 6.16.

5 Refer to Figure 6.17. Design an analog delay line that employs a two phase clock?

6.9 References

Baker, L. (1984), "Dynamic Transfer Networks", *IEEE Circuits and Systems*, Vol. 6, No. 4, pp. 24–6.

Bermudez, J. C. M. and Bhattacharyya, B. B. (1982), "Parasitic Insensitive Toggle Switched, Capacitor and its Applications to Switched-Capacitor Networks", *Electronic Letters*, Vol. 18, No. 17, 19 August, pp. 734–6.

Caves, J. T., Copeland, M. A., Rahim, C. F. and Rosenbaum, S. D. (1977), "Sampled Analogue Filtering Using Switched Capacitors as Resistor Equivalents", *IEEE Journal of Solid State Circuits*, Vol. SC–12, pp. 592–9.

Faruque, S. M., Vlach, M., Vlach, J., Singhal, K. and Viswanathan, T. R. (1982), "FDNR Switched Capacitor Filters Insensitive to Parasitic Capacitances", *IEEE Transactions on Circuits and Systems*, Vol. CAS–29, No. 9, pp. 589–95.

Fischer, J. H. (1982), "Noise Source and Calculation Techniques for Switched Capacitor Filters", *IEEE Journal of Solid State Circuits*, Vol. SC–17, No. 4, pp. 742–52.

Fleischer, P. E. and Laker, K. R. (1979), "A Family of Active Switched Capacitor Biquad Building Blocks", *Bell Systems Technical Journal*, Vol. 58, No. 10, pp. 2235–69.

Fudim, E. V. (1984), "Fundamentals of Switched-Capacitor Approach to Circuit Synthesis", *IEEE Circuits and Systems*, Vol. 6, No. 4, pp. 12–21.

Ghausi, M. S. (1984), "Analogue Active Filters", *IEEE Transactions on Circuits and Systems*, Vol. CAS-31, No. 1, pp. 13–31.

Gobet, C. A. and Knob, A. (1983), "Noise Analysis of Switched Capacitor Networks", *IEEE Transactions on Circuits and Systems*, Vol. CAS-30, No. 1, pp. 37–43.

Gregorian, R. (1982), "An offset-free Switched Capacitor Biquad", *Microelectronics Journal*, Vol. 13, No. 4, pp. 37–40.

Gregorian, R., Martin, K. W. and Temes, G. C. (1983), "Switched Capacitor Circuit Design", *Proc. IEEE*, Vol. 71, No. 8, pp. 941–66.

Herbst, D. and Hosticka, B. J. (1980), "Novel Bottom Plate Stray Insensitive Voltage Inverter Switch", *Electronic Letters*, Vol. 16, No. 6, 31 July, pp 636–7.

Howes, M. J. and Morgan, D. V. (1979), *Charge Coupled Devices and Systems*, Wiley, New York.

McCharles, R. H. and Hodges, D. A. (1978), "Charge Circuits for Analogue LSI", *IEEE Transactions Circuits and Systems*, Vol. CAS-25, pp. 490–7.

Mavor, J., Reekie, H. M. Denyer, P. B., Scanlan, S. O., Curran, T. M. and Farrag, A. (1981), "A Prototype Switched Capacitor Voltage Wave Filter Realized in nMOS Technology", *IEEE Journal of Solid State Circuits*, Vol. SC–16, No. 6, pp. 716–23.

Raut, R. and Bhattacharyya, B. B. (1984), "Designs of Parasitic Tolerant Switched Capacitor Filters Using Unity Gain Buffers", *IEEE Proc.*, Vol. 131, Part G, No. 3, pp. 103–13.

Saraswat, R. C. and Saha, S. K. (1984), "On the Design of Switched Capacitor Filters Using Cascade Sections", *International Journal of Electronics*, Vol. 57, No. 1, pp. 147–54.

Sasikumar, M., Radhakrishna Rao and Reddy, M. A. (1983), "Active Compensation in the Switched Capacitor Biquad", *Proc. of IEEE*, Vol. 71, No. 8, pp. 1008–9.

Temes, G. C. (1984), "Improved Offset Compensation Schemes for Switched Capacitor Circuits", *Electronic Letters*, Vol. 20, No. 12, 7 June, pp. 508–9.

Tsividis, Y. P. (1980), "Method for Signal Processing with Transfer Function Coefficients Dependent Only on Timing", *Electronic Letters*, Vol. 16, No. 21, 9 October, pp. 796–8.

Yasumoto, M. and Enomoto, T. (1982), "Integrated MOS Four-Quadrant Analogue Multiplier Using Switched Capacitor Techniques", *Electronic Letters*, Vol. 18, No. 18, 2 September, pp. 769–71.

7 Conversion methods

7.1 Introduction

Analog to digital and digital to analog converters are essential building blocks in many digital systems, allowing direct interfacing to the real analog world. Such converters have a number of important parameters which must be matched to the requirements of the overall system. The parameters include:

- monotonicity;
- resolution;
- offset; and
- settling time.

These in turn relate to:

- the specific method of conversion employed;
- the number of binary bits used;
- accuracy of matching components;
- effects of parasitic components;
- the frequency response of components;
- stability of the reference voltage/current; and
- the magnitude of the noise and signal voltages.

If we consider the conversion range possible with a MOS process, then the smallest signal levels are set by noise levels and comparator sensitivity. These may limit signals to being greater than a millivolt. The largest voltage that can be accommodated is set by the process breakdown voltage, which could be as high as 15 volts. Within these limits, (15 000–1) steps each of 1 millivolt (or a word length of 13–14 binary bits), could be accommodated. For normal multiproject chip work, this upper limit is impossible to achieve. Comparator sensitivities are typically 4 or 5 millivolts and, working from a 5 volt rail with analog circuits, a 2 volt swing is common. Thus, because only 400 or 500 steps are available, word lengths can only be eight or nine bits long.

An additional factor that is important at this stage is the matching of components. In Chapter 2, it was indicated that MOS transistors could be matched to about 5 percent, resistors $\frac{1}{2}$ to 1 percent and capacitors $\frac{1}{2}$ percent or better. It is this factor that ultimately limits the word length to about eight bits.

7.2 Digitial to analog converter types

Converters suitable for manufacture on a MOS process can be classified into several types (as shown in Figure 7.1): resistor, capacitor or transistor network, and combinations of these (Jain and Haskard, 1984). A number of well-known conventional methods are not applicable for the MOS process and are, therefore, omitted. Consider the binary weighted or resistive ladder networks. The network is made from n^+ diffusion or polysilicon where the sheet resistance is low, typically 20–100 ohm/square. On the other hand, the transistor active switches used to switch

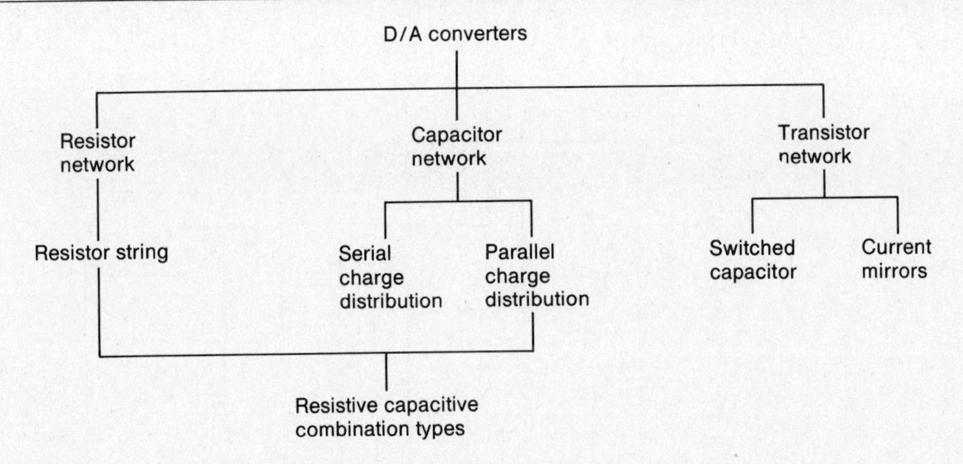

Figure 7.1 Digital to analog converter types that can be readily fabricated using the MOS process

resistors in and out of circuit have an ON sheet resistance of 10 kilo-ohm/square. Consequently, to prevent the switch from dominating the resistance, either the resistor values of the network must be made considerably greater than 10 k-ohm or the transistor switch must be designed with a large width to length ratio in order to reduce its ON resistance. In either case, a large amount of silicon is required and the response time of the converter can be poor because of large time constants, due either to the large network resistance or gate capacities of the transistor switches.

7.2.1 The resistor string converter

The simplest digital to analog converter to implement is the resistive string (shown in Figure 7.2). For an n-bit converter, the resistive string consists of a 2^n resistor string and $(2^n + 2^{n-1} + \ldots + 2^1)$ transistors. A reference voltage is fed onto one end of the string and, depending on the digital input, the voltage from the appropriate tap is switched through to the output. The magnitude of the reference voltage must always be less than the gate voltage by at least the threshold voltage of the enhancement transistor switches to ensure that there is no voltage drop across the switch.

Since the source voltage of these transistors can be at potentials approaching the reference voltage, the back gate voltage effect will significantly increase the threshold voltage. It is not uncommon in this situation for the threshold voltage to rise to 1.5 or even 2 volts. Consequently, for digital circuits feeding the enhancement transistors from a 5 volt supply rail, the reference voltage should be less than $2\frac{1}{2}$ volts. A problem with this type of converter is that there can be a large current flow through the string. Reducing the reference voltage magnitude decreases the current, but making the reference too small can cause other difficulties, particularly if the converter is part of an analog to digital converter where the comparator may no longer be able to detect a change of one bit. Practical reference voltages are usually between 1 and 2 volts.

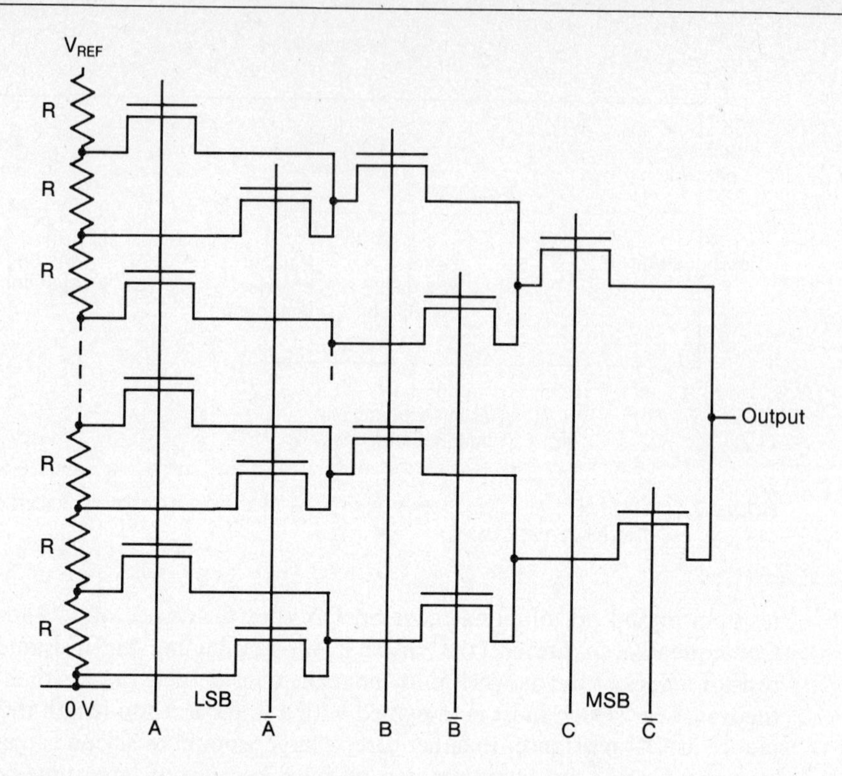

Figure 7.2 A three bit resistive string digital to analog converter

With the resistor string converter, it is very easy to change the weighting of the resistors so that logarithmic and other responses are possible rather than the linear case discussed here.

The layout of the string is most easily accomplished by using n^+ diffusion for the resistors, as this easily flows in to form the source/drains of the switching transistors. The layout can be easily parameterized to give a layout for any specified bit length. Because of resistor tolerances ($\frac{1}{2}$–1 percent), the maximum number of bits must be restricted to seven or less. Concern has been expressed by Huang (1980) over the resistance of the cuts at the ends of the string; care must be taken to ensure that they are small compared with the resistance of a string resistor. The layout for a resistive string converter is given in section 7.6.

Finally, should this digital to analog converter be employed as part of an analog to digital converter, it is preferable to feed into the converter comparator voltage levels shifted by half the amount of one bit. This offsets the threshold point to the centre of each increment. Consequently, the bottom resistor is halved and the top resistor is increased by 50 percent (as shown in Figure 7.3).

7.2.2 Serial charge redistribution converter

While a serial charge redistribution converter (Suarex et al., 1975) is simple conceptually, it is very difficult to implement and to retain accuracy. As shown in

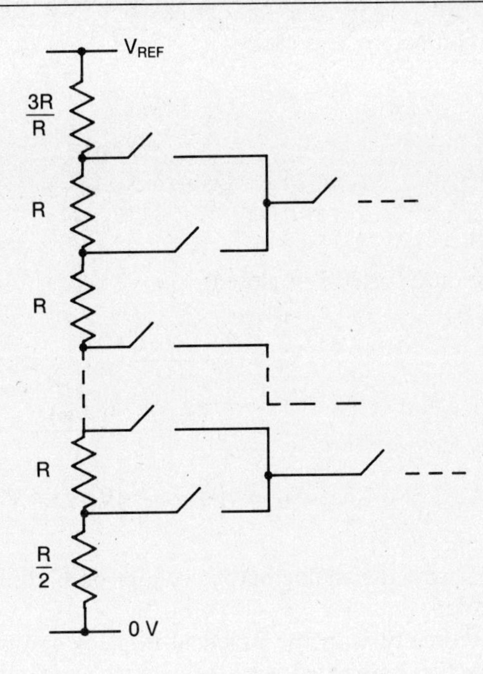

Figure 7.3 Modified resistor string converter for inclusion in an analog to digital converter

Figure 7.4, the circuit consists of two identical capacitors ($C_1 = C_2 = C$). Initially, the two capacitors are reset by short circuiting them with switches S_4 and S_5. Next, the digital input is applied in serial form, commencing with the least significant bit a_0. If a_0 is a "1", S_1 closes, applying V_{REF} to S_2. If a_0 is a "0", zero volts is applied to S_1. While a_0 is still applied, S_2 closes, so that C_1 charges to a value a_0V_{REF}. Switch S_2 opens and S_3 closes, with the charge on C_1 being redistributed or shared with C_2, the resulting voltage left on C_2 when S_3 opens being $\frac{1}{2}a_0V_{REF}$. The cycle recommences with the next to least significant bit, a_1, being applied so that C_1 is charged up to a_1V_{REF}. When S_2 opens, S_3 closes, the total shared charge on C_1 and C_2 being as follows:

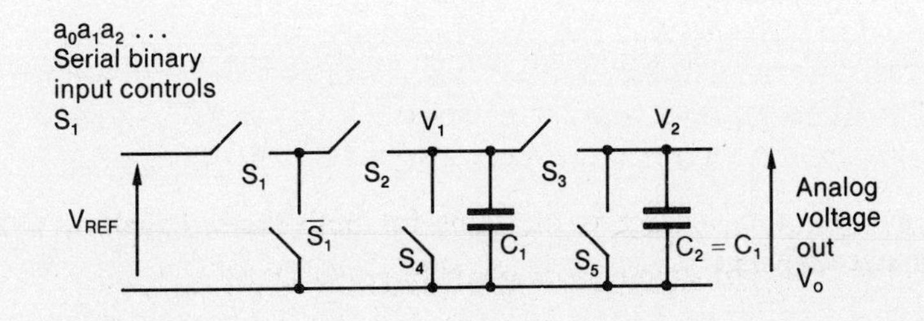

Figure 7.4 Serial charge redistribution digital to analog converter

Initial condition before S_3 is closed:

$$\begin{aligned} Q_1 &= V_1C_1 \\ &= a_1V_{REF}C \\ Q_2 &= V_2C_2 \\ &= \tfrac{1}{2}a_0V_{REF}C \end{aligned}$$

Final voltage on C_2 after S_3 is closed:

$$\begin{aligned} V_2 &= \frac{Q_1 + Q_2}{C_1 + C_2} \\ &= \frac{CV_{REF}(a_1 + \frac{1}{2}a_0)}{2C} \end{aligned}$$

$$V_2 = \left(\frac{a_1}{2} + \frac{a_0}{4}\right)V_{REF} = V_0 \tag{7.1}$$

In this way, the analog output voltage on C_2 builds up in the required binary weighted fashion.

The difficulty with the practical implementation of the circuit is the stray junction capacities associated with the sources and drains of the switches. Not only are the capacitors large, they are also very voltage dependent, resulting in errors of charge and consequently errors in the output analog voltage. To overcome this problem, capacitors C_1 and C_2 must be made large in area, with the circuit losing its initial apparent attractiveness of occupying a small area, by requiring only two capacitors.

7.2.3 Parallel charge redistribution

The principle of the parallel charge redistribution converter (see McCreary and Gray, 1975; Albarran and Hodges, 1976; Gregorian, 1981) is illustrated in Figure 7.5. A two-phase nonoverlapping clock is used. During phase 1, the binary weighted capacitors are switched to the reference or zero voltage, depending on the parallel digital inputs. The total charge on all capacitors (for the three bit case given in the figure) is:

$$Q_{total} = CV_{REF}(4a_2 + 2a_1 + a_0) \tag{7.2}$$

On phase 2, all capacitors are shorted together, the resulting charge being:

$$Q_{total} = 7CV_0 \tag{7.3}$$

Since the total charge is the same for both cases, equations 7.2 and 7.3 can be equated, giving:

$$V_0 = \frac{V_{REF}}{7}(4a_2 + 2a_1 + a_0) \tag{7.4}$$

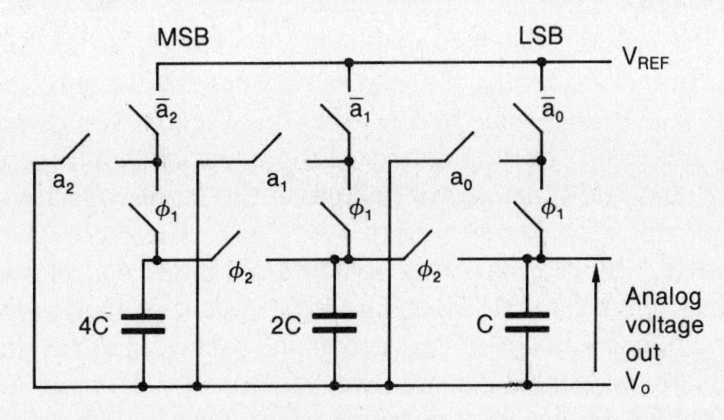

Figure 7.5 A three bit parallel charge redistribution digital to analog circuit

The seven in the denominator is inconvenient but it can be increased to eight thereby cancelling out if a dummy capacity of value C is added to the circuit. During phase 1, the new capacity is short circuited and, on phase 2, switched in parallel with the other capacitors. As a result, equation 7.2 remains unchanged while equation 7.3 becomes:

$$Q_{total} = 8CV_0$$

giving the analog output voltage of:

$$V_0 = V_{REF}\left(\frac{a_2}{2} + \frac{a_1}{4} + \frac{a_0}{8}\right) \tag{7.5}$$

We will see later that this additional capacitor can be put to good use.

At first, it would seem that the parallel charge redistribution circuit has the same problem as the serial case, that is, inaccuracies due to the switch parasitic capacities. While this is true of the circuit in its present form, the parallel case can be rearranged to place all of the switches on the earthed side of the capacitors (as shown in Figure 7.6). The extra dummy capacitor has also been included.

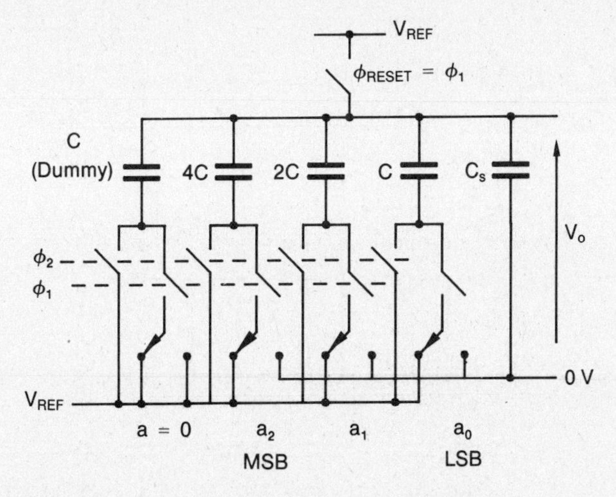

Figure 7.6 Inverted parallel switched capacitor converter

The circuit still has one switch connected to the output used to reset the system on ϕ_1. To reduce the effects of its source to substrate capacitance, reset is made to V_{REF} rather than zero volts so that the junction reverse voltage is increased, thereby reducing the junction parasitic capacity (see Figure 2.17). This junction capacity and any other strays (including the input capacity of the amplifier or any other circuit the converter may feed) have been grouped together as a single capacitor C_s. Since ϕ_{RESET} now switches the upper end of the capacitors to V_{REF}, if the digital input is a "1", then the lower plates of capacitors are switched to zero volts to apply a voltage of V_{REF} across the capacitor. If the digital input is a "0", the lower plates of capacitors are switched to V_{REF} to place zero voltage across the capacitor. The output voltage during phase 2 clock is, therefore,

$$V_0 = V_{REF}\left[1 + \left(\frac{1}{2}a_2 + \frac{1}{4}a_1 + \frac{1}{8}a_0 + \frac{C_s}{8C}\right)\right] \tag{7.6}$$

Equation 7.6 is of the same form as 7.5 except there is the offset of V_{REF} and an error term V_{REF} $C_s/8C$. For this error term to be insignificant, we require $C_s/C \ll a_0$. In practice, C_s is determined, and C is chosen so that $0.1 < C_s/C < 0.25$.

Where feedback techniques are employed around an amplifier, it is possible to hold the voltage across C_s constant so that the effect of the stray capacities is removed altogether. Consider the converter given in Figure 7.7. The feedback capacitor has been made a multiple N of the unit binary weighted capacitors.

During phase 1, the amplifier is biased to its self bias voltage of V_B (V_O is zero). If the input is a logical "1", the binary weighted capacitors are switched to V_{REF}; if the input is a logical "0", the capacitors are switched to zero volts. On phase

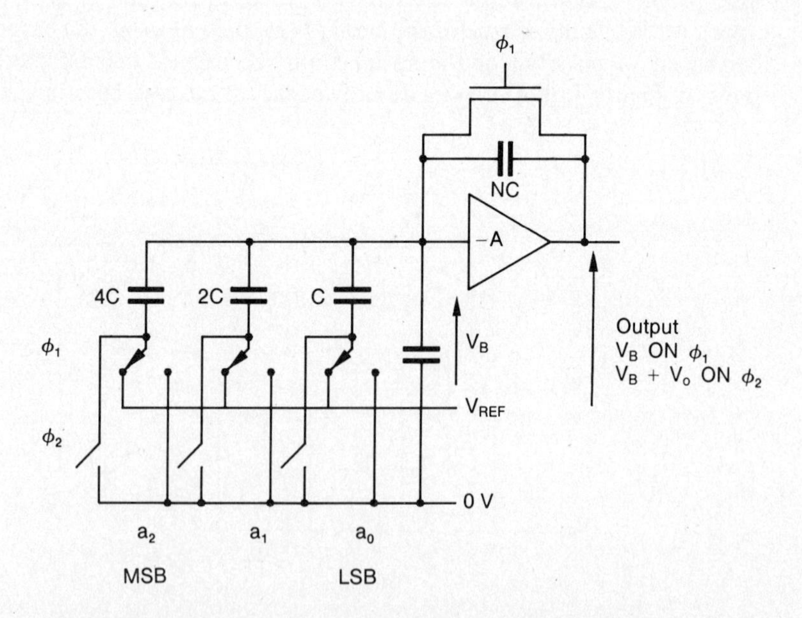

Figure 7.7 Improved charge distribution digital to analog converter

2, the capacitors are all switched to zero volts. As a result, the charge on the capacitors during phase 1 is as follows:

$$Q_{total} = 4C(V_B - a_2V_{REF}) + 2C(V_B - a_1V_{REF}) + C(V_B - a_0V_{REF}) + C_sV_B \qquad (7.7)$$

During phase 2, the output of the amplifier changes and the capacitor charge flows into the feedback capacitor so that,

$$Q_{total} = 4CV_B + 2CV_B + CV_B + NC[V_B - (V_0 + V_B)] + C_sV_B \qquad (7.8)$$

Equating these equations gives,

$$V_0 = V_{REF}\left(\frac{4}{N}a_2 + \frac{2}{N}a_1 + \frac{a_0}{N}\right) \qquad (7.9)$$

It has been assumed in this derivation that the magnitude of the amplifier gain A is large so that on phase 2 the change in amplifier input voltage (V_0/A) can be ignored. Thus the voltage across C_s is constant and no error is introduced.

Note that the output voltage V_0 is, with respect to the bias voltage, V_B. This is useful as it allows coupling into other self biased stages.

The value of the feedback capacitor can be used to control the converter gain. If N equals eight, then equation 7.9 is of the same form as the original equation 7.5. Should N be made unity, the converter would not only occupy less area but it would also have a voltage gain of eight.

Finally, because capacitors can be matched to better than $\frac{1}{2}$ percent, it is possible to have converters of up to eight bits. Unfortunately, this is not always desirable as such a binary weighted capacitor system would occupy a very large area. One way to reduce the size of the area is to break the capacitors into subgroups of not more than four bits and use the gain factors (referred to in equation 7.9) to scale groups when combining.

7.2.4 Combined resistor string and charge redistribution

A practical solution to the dual dilemma of eight-bit resistor string converters not having equal steps because of poor resistor matching and eight-bit charge redistribution types occupying excessive area is to combine the two systems. There are a number of ways of doing this (Fotouhi and Hodges, 1979; Sherman, 1984). We will consider a number of alternatives and as an example use a five-bit resistor string (Figure 7.2) combined with a three-bit charge redistribution circuit (Figure 7.5) to make an eight-bit converter. With the charge redistribution circuit, an additional dummy capacitor of unit value C was previously added so that cancellation occurred. We can now use this capacitor to feed in the output from our resistor string. Consider the circuit in Figure 7.8. Again ϕ_1 and ϕ_2 are two phase nonoverlapping clocks. During phase one, the binary weighted capacitors are charged to V_{REF} or zero, depending on the digital inputs. The dummy capacitor,

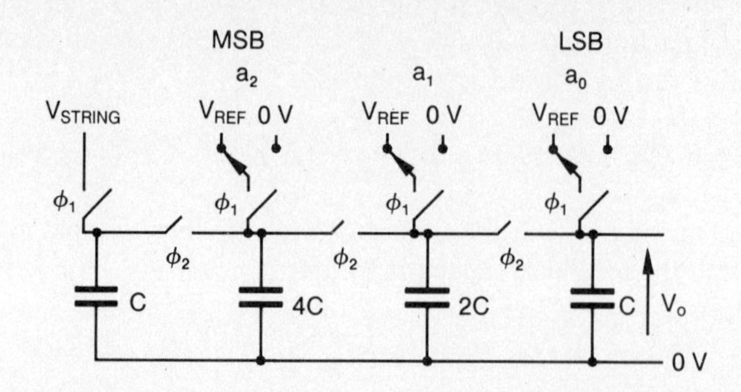

Figure 7.8 Combined resistor string and charge redistribution converter

however, is always charged to the voltage, out of the resistor string. Therefore, for digital inputs to the string of b_0 to b_4,

$$V_{STRING} = V_{REF}\left(\frac{b_4}{2} + \frac{b_3}{4} + \frac{b_2}{8} + \frac{b_1}{16} + \frac{b_0}{32}\right) \quad \textbf{(7.10)}$$

The charge stored on the capacitors will be,

$$Q_{total} = 4Ca_2V_{REF} + 2Ca_1V_{REF} + Ca_0V_{REF} + CV_{STRING} \quad \textbf{(7.11)}$$

During phase 2, all capacitors are switched in parallel to give an output voltage V_0 where,

$$Q_{total} = 8CV_0 \quad \textbf{(7.12)}$$

Combining equations 7.10 to 7.12, we obtain,

$$V_0 = V_{REF}\left(\frac{a_2}{2} + \frac{a_1}{4} + \frac{a_0}{8} + \frac{b_4}{16} + \frac{b_3}{32} \ldots + \frac{b_0}{256}\right) \quad \textbf{(7.13)}$$

which is what is required for an eight bit converter.

The improvements made to the charge distribution circuit in section 7.2.3 can also be accommodated here. For example, if the switches are placed at the lower end of the capacitor, as in Figure 7.6, then the output voltage from the resistor string must be complemented for it to be consistent with the capacitor circuit. This is achieved simply by reversing the voltages at either end of the string (Figure 7.2). Thus, for the input of 00000 into the string, the output voltage is now V_{REF}.

Thus far, we have explored only one method of combining the two circuits of Figure 7.8. Another method could be to make use of the tapped voltages available from a resistor string network. What we wish to do is to binary weight charges. We have achieved this by using a single voltage V_{REF} and changed the size of the capacitors. Alternatively, we could have kept the capacitors the same value and used binary weighted voltages obtained from a resistor string. Figure 7.9 shows two possible arrangements. However, both circuits have at least two disadvantages

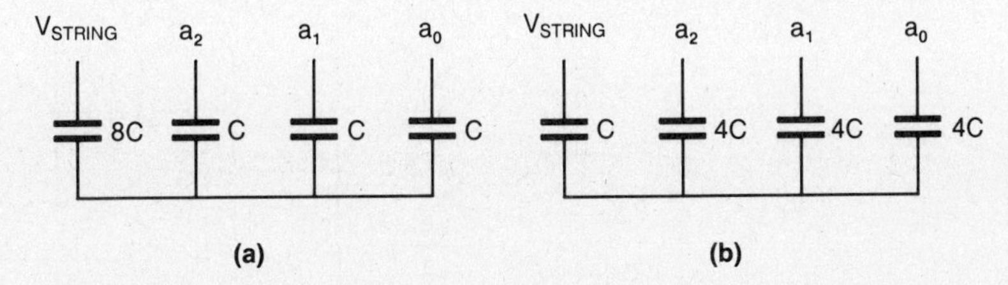

Figure 7.9 Alternative methods of combining a resistor string and a charge redistribution converter

when compared with the previous design of Figure 7.8. First, they require additional routing, for three reference voltages, V_{REF}, $\frac{V_{REF}}{2}$ and $\frac{V_{REF}}{4}$, must be routed.

More importantly, the two circuits occupy additional area, for circuit (a) has a total of 11C and circuit (b) has 13C, while the original circuit has a total capacity of only 8C.

Using this combined capacitor resistor string technique, it is possible to fabricate a digital to analog converter of eight bits. Section 7.5 provides further details.

7.2.5 Switched capacitor type

We saw in Chapter 6 that if a capacitor is switched at a rate much faster than the maximum input frequency, the circuit can be represented by an equivalent resistor, thus,

$$R_{EQ} = \frac{1}{fC} \tag{7.14}$$

where f is the frequency at which the capacitor is switched and C is the value of the capacitor. The value of this equivalent resistor (representing the amount of charge being transferred) can be increased by either reducing C or the frequency. We now have a means of producing a binary weighted switched capacitor system where we can either binary weight the capacitors or, by using binary counters, halve the switching frequency.

The principle is illustrated in Figure 7.10. In both cases,

$$V_0 = -V_{REF}\left(a_2 + \frac{a_1}{2} + \frac{a_0}{4}\right) \tag{7.15}$$

Again MOS tiles are available in the appendix which readily allow such circuits to be fabricated. Circuit (a) of Figure 7.10 is the one that is most usual, for while it has a larger capacity area, there are fewer problems in clock generation and distribution.

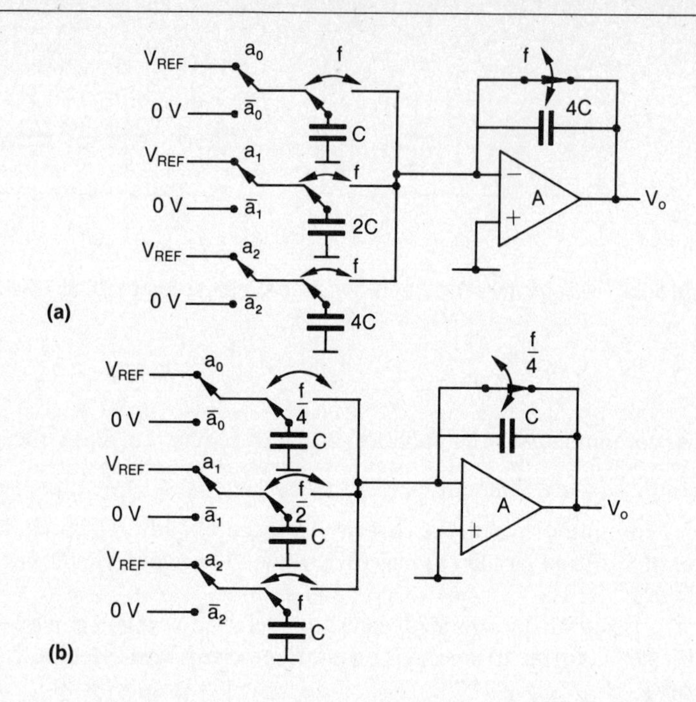

Figure 7.10 Examples of switched capacitor digital to analog converters

7.2.6 Transistor current mirrors

The traditional method of making a converter by using a linear bipolar process is to generate binary weighted currents using current mirrors. The same technique can be employed with MOS enhancement or lateral bipolar transistors (Yee, 1978). Consider the simple current mirror circuit given in Figure 7.11. Parallel transistors are used in the mirrors to achieve good current matching and a high degree of replication. A push pull system is used so that current flows continually in the mirrors. By using the digital switching as shown, the output V_0 is positive going as the binary input increases. By interchanging the logical input signal and its complement, the output voltage can be made to decrease with an increasing binary input. With the example given in Figure 7.11, for an input of 000 all the current flows into the resistor on the positive input of the amplifier, giving across this resistor a voltage of,

$$\begin{aligned} V_R &= -IR(4a_2 + 2a_1 + a_0) \\ &= V_0 \end{aligned} \qquad (7.16)$$

Since the amplifier has 100 percent feedback, the output voltage V_0 is equal to V_R (as given in equation 7.16). At the upper end of the binary input, namely 111, all current flows into the feedback resistor giving,

$$V_0 = +IR(4a_2 + 2a_1 + a_0) \qquad (7.17)$$

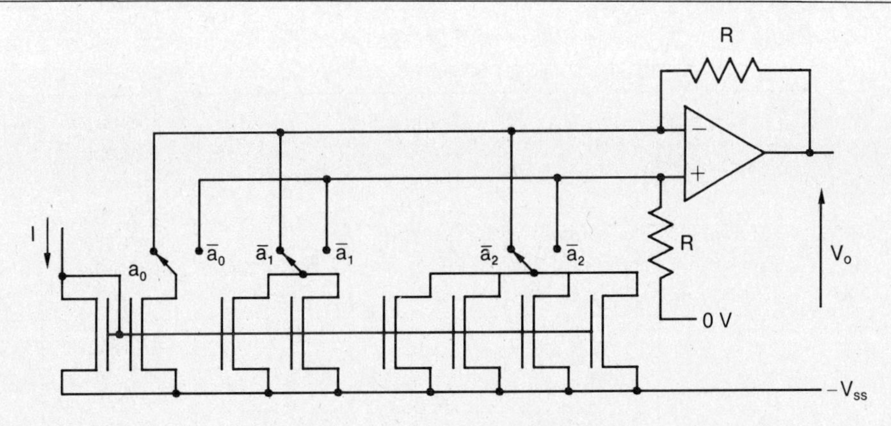

Figure 7.11 Weighted current digital to analog converter

The difficulty experienced with this method is as a result of variation of currents in the mirrors due to drain source or collector emitter voltage changes. This can be considerably reduced by employing the three transistor Wilson current mirrors (as shown in Figure 7.12). Here the voltages across the two mirror transistors M1 and M2 have drain source voltages which are kept in the same order. There is a more critical difficulty of the accuracy to which the current mirror transistors can be matched. At low currents (one normally endeavors to keep current levels and therefore dissipation of the chip as low as possible), MOS transistors are operating near their threshold voltages and are therefore very dependent on any surface charge. Fortunately, this is not the case with the bipolar transistors, in which case better matching can be obtained at low currents.

Table 7.1 gives measurements made on a simple four bit converter identical to Figure 7.11, except that Wilson current mirrors were employed. The table gives the case for 1111 input so that the output current from the mirror should be 15 times the input current. Low current matching of MOS transistors is poor and it is only when one has moved well away from the threshold point, and currents become large, that matching starts to be acceptable. With bipolar transistors this is not the case.

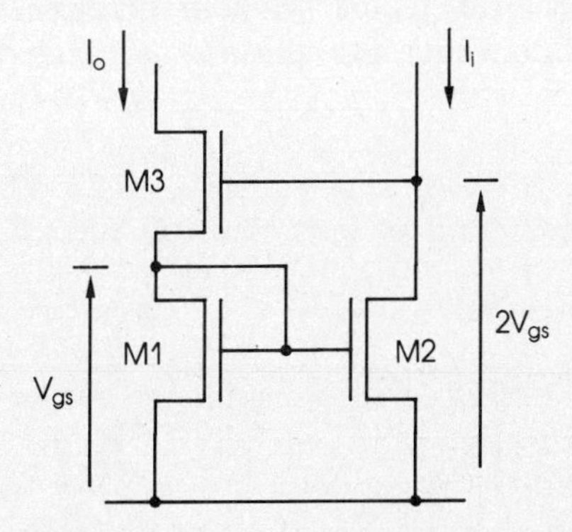

Figure 7.12 Wilson current mirrors for reducing drain source voltage variation effects

Table 7.1 Calculated ideal and actual (measured) circuits for Wilson current mirror digital to analog converters, using nMOS and lateral pnp bipolar transistors

Input current to mirror μA	*Ideal output current (× 15)* μA	*Measured output current nMOS transistor converter* μA	*Percentage error*	*Measured output current bipolar transistor converter* μA	*Percentage error*
0.5	7.5	—	—	7.542	+0.56
1	15	—	—	15.29	+1.9
3	45	48.65	+8.1	47.07	+4.6
5	75	79.6	+6.1	79.8	+6.4
10	150	156.9	+4.6	—	—
15	225	231.0	+2.7	—	—
20	300	305.0	+1.6	—	—

7.3 Analog to digital converters

7.3.1 Introduction

Analog to digital conversion is generally more difficult to achieve than the digital to analog, as additional circuitry is usually required. These converters can be categorized under the three headings shown in Figure 7.13.

7.3.2 A/D converters employing D/A converters

This type of converter is often preferred as the majority of the additional circuitry is digital and therefore easily implemented in MOS technology. The additional analog circuit is a comparator which will be discussed separately. Consider the simpler staircase converter shown in Figure 7.14.

An n-bit counter (n being the converter word length) is incremented and its output converted to an analog signal as well as compared with the unknown analog input signal. Once the counter has been incremented to the point where the output from the D/A converter just exceeds the analog input, the comparator output

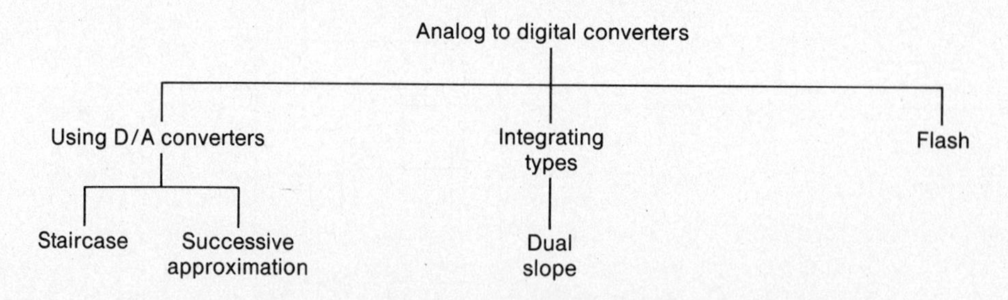

Figure 7.13 Classification of analog to digital converters

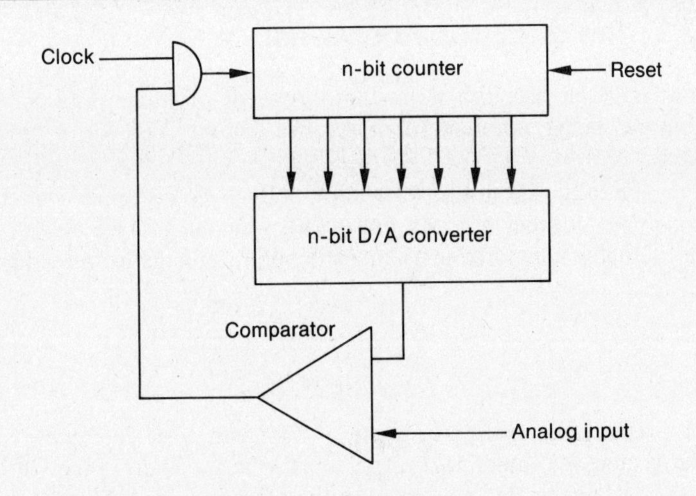

Figure 7.14 Staircase analog to digital converter

switches, preventing the clock further incrementing the counter. The counter contains the binary representation of the analog input voltage. Perhaps the major disadvantage of this simple system is that the conversion time is variable and can take up to 2^n clock pulses, depending on the input voltage.

The successive approximation method, as shown in Figure 7.15, is fast and takes a constant time to achieve a conversion. The ring counter of n + 2 bits length initially sets the most significant bit (MSB) of the output register. Through the D/A converter, this is compared with the analog input. If it is too great, then on the next clock pulse the control module not only sets the second to MSB to one, but resets the MSB. If the MSB was insufficient, it is not reset on the next clock pulse. Thus, with n + 2 clock pulses, the output register is set up to the binary equivalent of the analog input signal.

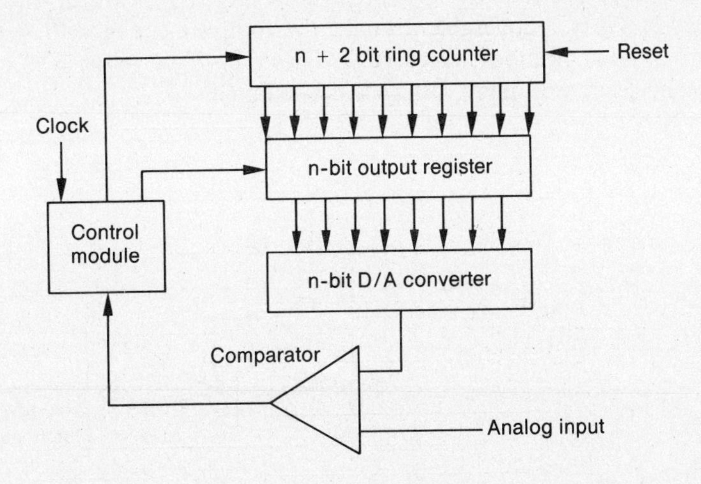

Figure 7.15 Successive approximation analog to digital converter

7.3.3 The dual slope converter

This converter has the important property of integrating the input signal and, therefore, being immune to noise. See Figure 7.16 for a block diagram of the system.

Initially, the unknown analog voltage (V_i) is applied to the integrator input for a preset number of clock pulses (n). Thus, if T is the clock period, the time the input signal is integrated is nT and the output voltage from the integrator is,

$$-V_{INT} = \frac{nTV_i}{C} \tag{7.18}$$

Next, the integrator input is switched to a reference voltage of opposite polarity and the integrator output falls to zero. The number of clock pulses taken for this to happen (m) is counted in the output register. Thus,

$$V_{INT} = \frac{mT(-V_{REF})}{C} \tag{7.19}$$

Equating equations 7.18 and 7.19, we obtain,

$$m = \frac{nV_i}{V_{REF}} \tag{7.20}$$

The output voltage register displays a value proportional to the input voltage. With the correct selection of n/V_{REF}, the scaling can be made any convenient factor, with unity being usual.

For nMOS technology, the difficulty with this circuit is the analog integrator circuits. Circuits discussed in Chapter 6 can be used. (See Figure 7.17 for an example of one such circuit.) Because signal voltages are all with respect to the bias voltage V_B, it is preferable to make the complete circuit with respect to this voltage rather than common or zero volts. Thus the $-V_{REF}$ voltage can be made zero and a second reference supply voltage is not required.

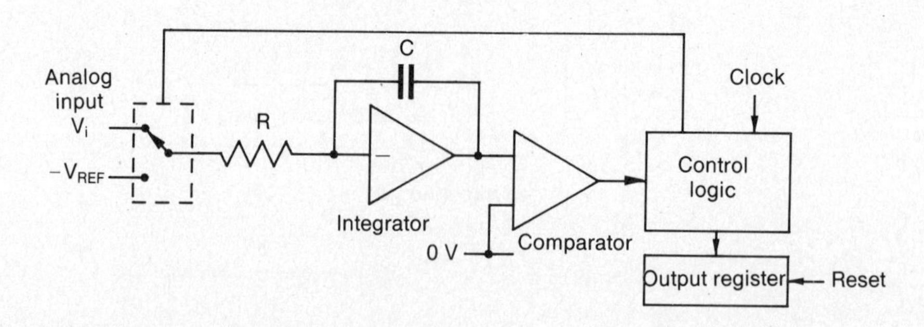

Figure 7.16 Dual slope A/D converter

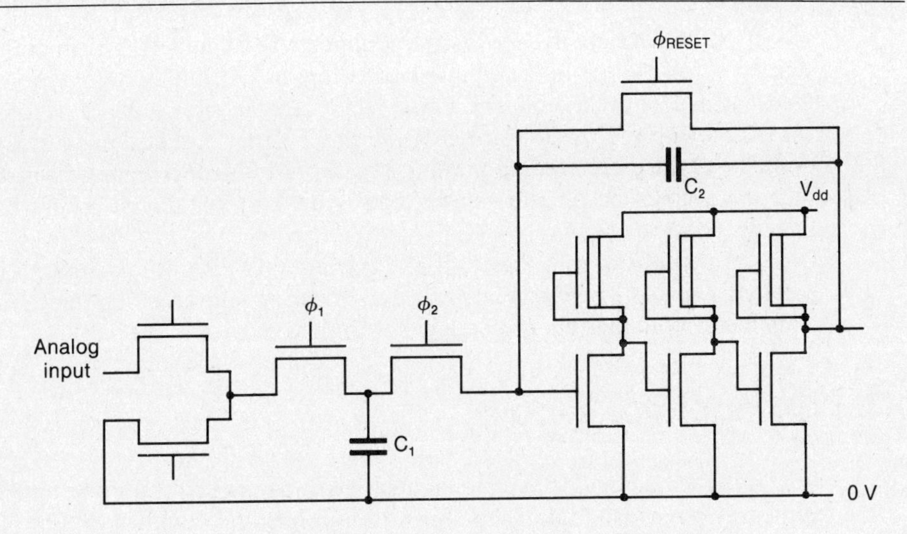

Figure 7.17 Simple integrator input circuit for the dual slope converter

7.3.4 Flash converters

The flash converter is the fastest of all converters and consists (as shown in Figure 7.18) of a series of comparators, with reference voltages generated from a resistance string. Simple logic converts the output of the comparators into the binary equivalent of the analog input. For an n-bit converter, 2^{n-1} comparators are required. The resistor string is also offset in that the top and bottom resistors are modified to place the threshold voltage in the centre of an increment.

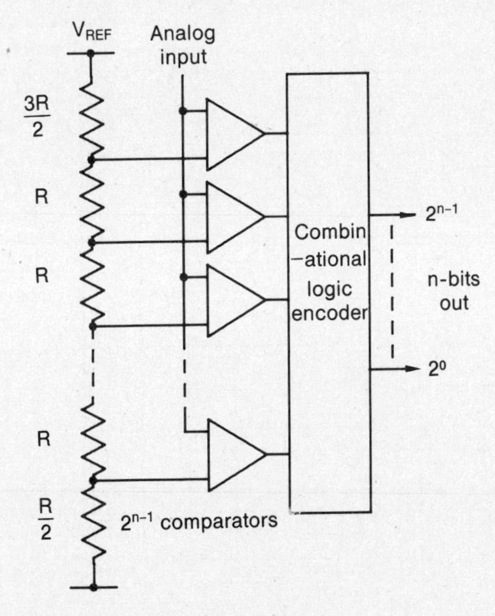

Figure 7.18 A flash analog to digital converter

If the ultimate in speed is not required, the circuit can be simplified by time sharing comparators. In the simplest case (Figure 7.19), only a single comparator is used, with a conventional two phase shift register circulating a 1 to switch the comparator up the resistor string. If the shift register is replaced by a counter with decoder, or the clock feeding the shift register also drives a separate binary counter, then the encoder becomes a simple "and" gate, with the digital output being taken directly from the counter.

The problem with these types of converters is resistor matching. As discussed previously, the $\frac{1}{2}$–1 percent matching prevents these converters on multiproject chips from being more than six or seven bits word length.

7.4 Comparator circuits

In CMOS technology, conventional analog techniques can readily be applied, with a comparator normally taking the form of a differential amplifier or full operational amplifier, driving a latch (Allstot, 1982; Yukawa, 1985). This is not the case for nMOS.

As discussed in Chapter 5, the pseudoanalog comparator (Fotouhi and Hodges, 1979; Yee et al., 1978; Allstot, 1982) consists of either a switched capacitor input amplifier driving a bistable circuit (Figure 5.7, page 91) or possibly a modified bistable circuit by itself (Figure 5.11, page 94). A good example of the latter is the resistor string converter of Figure 7.19. Each time an input from the resistor string is fed in for comparison, the bistable is reset and a comparison is then made. Increasing the aspect ratio of the inverters used to make the bistable,

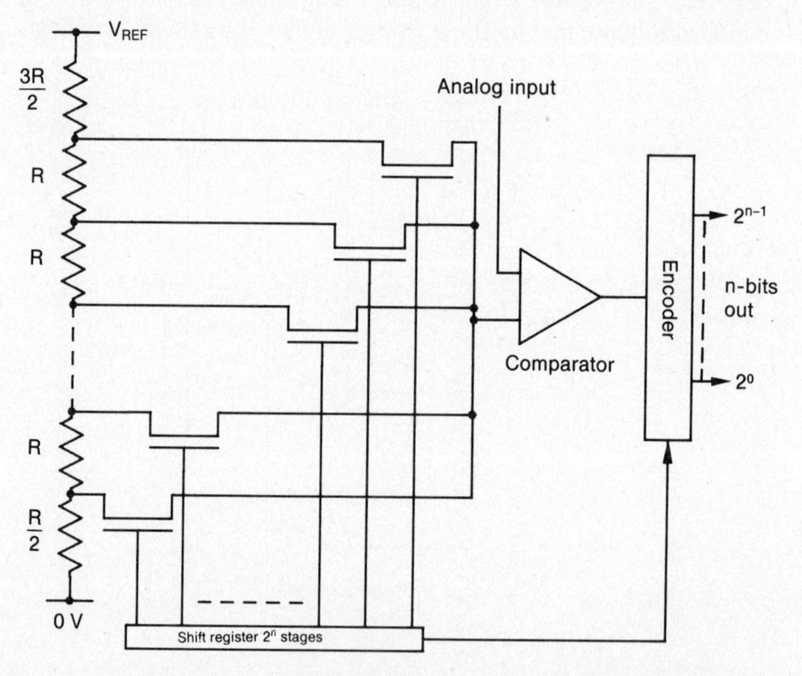

Figure 7.19 Simple resistor string A/D converter

improves the regenerative gain. Two 9:1 inverters give a loop gain of typically 200 and a sensitivity of approximately 10 millivolts.

For the majority of cases, the more complex switched capacitor circuit is required. It can be constructed from standard tiles and arranged so that the inputs for comparison can be accepted on either the same clock or opposite clock phase. Two typical circuits are shown in Figure 7.20. The parasitic associated with the input capacitor must be positioned so that is provides minimum attenuation. The sensitivity of this type of converter is typically 3 or 4 millivolts. (Sherman, 1984.)

7.5 The voltage reference

In bipolar technology, two methods can be employed to generate a constant reference voltage. The simplest method employs zener or avalanche breakdown in a

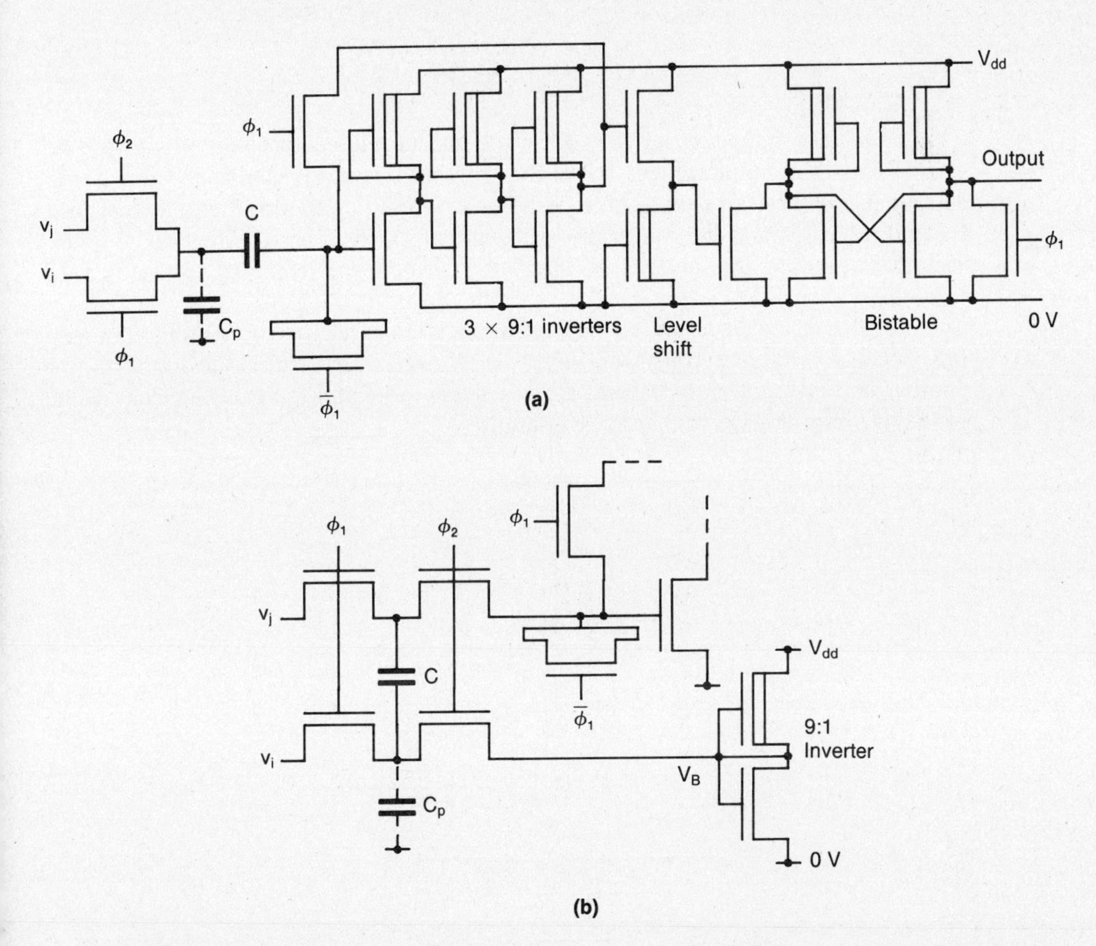

Note:
Circuit (a) requires inputs on opposite clock phases while (b) is a modification of the original circuit, enabling inputs to be on the same clock phase.

Figure 7.20 Switched capacitor voltage comparator circuits

reverse biased base emitter diode. This method has three difficulties: it is noisy, can have a large temperature coefficient, and the breakdown voltage is not only predetermined by the process but can vary from batch to batch. With MOS processes, this method is not practical; to ensure correct MOS operation diode breakdown voltages are always greater than the supply voltages employed.

The second method uses the band gap reference approach. Here there must be a forward biased diode. The band gap approach can be employed with the CMOS process but not with nMOS (Rehman, 1980), in which case the difference in threshold voltages between an enhancement and depletion transistor approach is used (Blanschild, 1978; Hoff, 1979; Haskard, 1983).

Perhaps the simplest circuit is a standard self biased inverter, as shown in Figure 7.21.

Using the simple model for both transistors in the saturation region (equation 2.13) and equating drain currents, we obtain the following:

$$V_{REF} = V_{to1} - V_{to2}\left(\frac{L_1}{L_2} \times \frac{W_2}{W_1}\right)^{\frac{1}{2}} \tag{7.21}$$

Because V_{to2} is a negative voltage, the output reference voltage will always be greater than the enhancement transistor threshold voltage, the final value being dependent on the transistor geometries. For example, if the threshold voltages are 0.7 and −3 volts, and the transistors are identical, the reference voltage is 3.7 volts. Should a standard 9:1 inverter be used, as expected, the reference voltage is only 1.7 volts.

Table 7.2 gives results for such a simple voltage reference. The difficulty with this circuit is that the transistor threshold voltages vary from one fabricator to another and even different batches from a single fabricator, so that the exact value of the reference voltage can never be guaranteed.

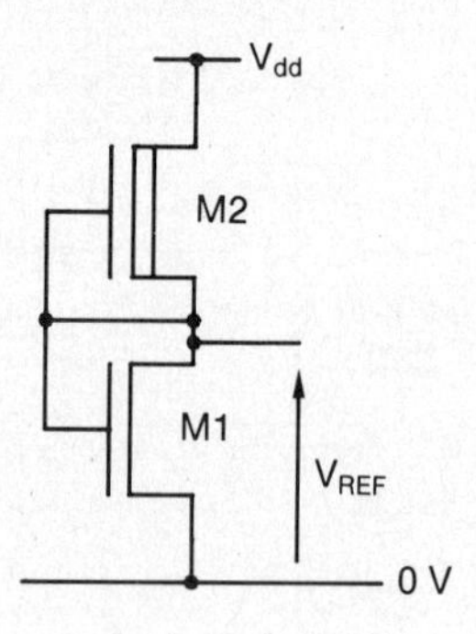

Figure 7.21 Simple inverter constant voltage source

Table 7.2 Output voltage against input voltage for a simple inverter reference

Input voltage volts V_{dd}	3	4	5	6	7
Output voltage volts V_{REF}	1.502	1.607	1.634	1.636	1.639

An improvement of this simple circuit, given in Figure 7.22, employs conventional analog techniques. The differential amplifier type circuit uses an enhancement transistor on one side and a depletion type on the other. Transistors M_3 and M_4 are of the same geometry as M_1 and M_2.

The output voltage V from this circuit is proportional to the difference in the threshold voltages of transistors M_1 and M_2, that is,

$$V = K(V_{t2} - V_{t1}) \quad \textbf{(7.22)}$$

where K is the proportionality constant. This is fed to an operational amplifier whose gain is set by resistors R_1 and R_2. Thus,

$$V_{REF} = \frac{R_1 + R_2}{R_2}(V_{t2} - V_{t1}) \quad \textbf{(7.23)}$$

While there will be production spreads in the term $(V_{t2} - V_{t1})$, if resistors R_1 and R_2 are external to the chip, the reference voltage can be adjusted to its desired value.

The final group of nMOS circuits is based on the modified standard inverter (shown in Figure 7.23), which is more advantageous than the previous types in that it has negative feedback to stabilize the output voltage and the ability to operate from low supply lines. Using the simple model given in equation 2.13, it can be shown that for the transistors in saturation,

$$V_{REF} = \frac{V_{to1} - \sqrt{\frac{W_2}{W_1} \times \frac{L_1}{L_2}}\, V_{to2}}{1 + \sqrt{\frac{W_2}{W_1} \times \frac{L_1}{L_2}}} \quad \textbf{(7.24)}$$

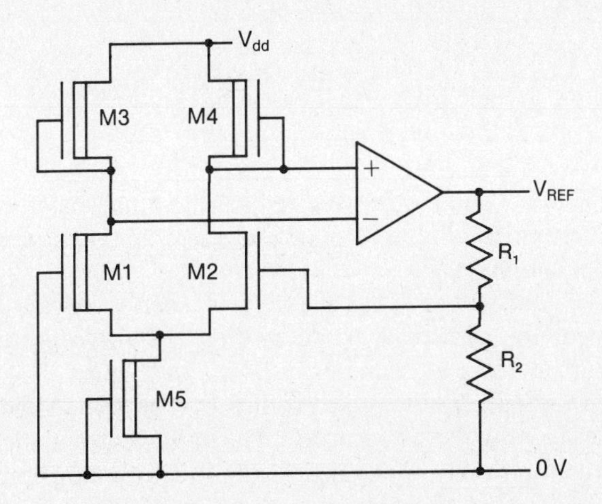

Figure 7.22 Improved voltage reference circuit

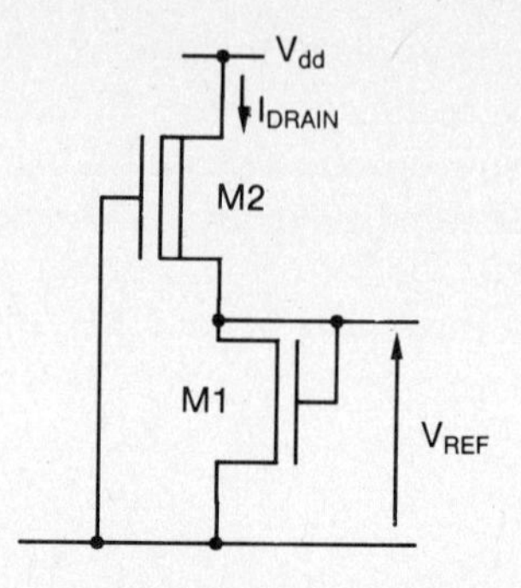

Figure 7.23 Simple constant voltage source

Again, assuming threshold voltages of 0.7 and −3 volts and identical geometry transistors, we obtain a reference voltage of 1.85 volts. Because the depletion transistor is biased by this voltage (negative feedback), it operates at a gate voltage much less than normal and is biased into a region where it behaves like a more ideal device, giving almost constant current characteristics with varying drain source voltage. We can use this to undertake a more exact analysis of the circuit. We can retain the simple model of equation 2.13 for M_2 but use the model of equation 2.19 for M_1 to allow for channel shortening effects. Equating drain currents of the two transistors gives

$$\frac{W_1}{L_1} \times \frac{L_2}{W_2}\left(\frac{V_{REF} - V_{t1}}{-V_{REF} - V_{t2}}\right)^2 = 1 - \lambda_1 V_{REF}$$

This is a cubic equation, but it has only one practical solution, namely,

$$V_{REF} = \frac{1}{\lambda_1}\Bigg\{-\left(1 + \sqrt{\frac{W_1L_2}{W_2L_1}} + \frac{\lambda_1 V_{t2}}{2}\right) + \left(1 + \sqrt{\frac{W_1L_2}{W_2L_1}} + \frac{\lambda_1 V_{t2}}{2}\right)^2 - 2\lambda_1\left(V_{t2} - \sqrt{\frac{W_1L_2}{W_2L_1}}\,V_{t1}\right)\Bigg\} \quad \textbf{(7.25)}$$

Table 7.3 summarizes the results of a small batch of such a regulator under no load current conditions. Notice that there are variations in the reference voltage value from chip to chip.

The temperature characteristic of such a circuit can be improved if the geometry of the devices is made so that the drain current is equal to the zero temperature coefficient value (as shown in Figure 7.24). This applies to both enhancement and depletion type transistors. The zero temperature coefficient current occurs because two effects cancel. The drain current increases with temperature because the threshold voltage decreases, also with temperature. However, the drain current will decrease with temperature, due to changes in carrier mobility. Above I_{d0}, the first effect is the dominant one, while below that current, the second effect

Table 7.3 Measured results for the modified inverter constant reference voltage circuit

Unregulated input voltage V_{dd}	*Unit 1*		*Unit 2*		*Unit 3*	
	V_{REF}	I_{drain}	V_{REF}	I_{drain}	V_{REF}	I_{drain}
3.0	1.727	7.7	1.763	8.4	1.774	8.3
4.0	1.737	8.0	1.775	8.7	1.786	8.6
5.0	1.744	8.2	1.781	8.9	1.791	8.8
6.0	1.748	8.4	1.786	9.0	1.796	9.0
8.0	1.755	8.7	1.793	9.4	1.803	9.3
10.0	1.761	9.0	1.800	9.7	1.809	9.7

Note:
For both transistors:
$L = 15\ \mu m$
$W = 10\ \mu m$
$V_{bs} = 0V$
$T = 23°C$

dominates. The current at which these effects cancel (Blanschild et al., 1978) is given as follows:

$$I_{d0} = \mu C_{ox} \frac{W}{L} k(\alpha T)^2 \tag{7.26}$$

where α is the gate threshold temperature coefficient and k is Boltzmans constant.

The value for I_{d0} varies from process to process but is typically in the range 1–10 μA. However, measured values have been as low as 0.3 μA. The temperature characteristic of Unit 1 of Table 7.3 is given in Table 7.4.

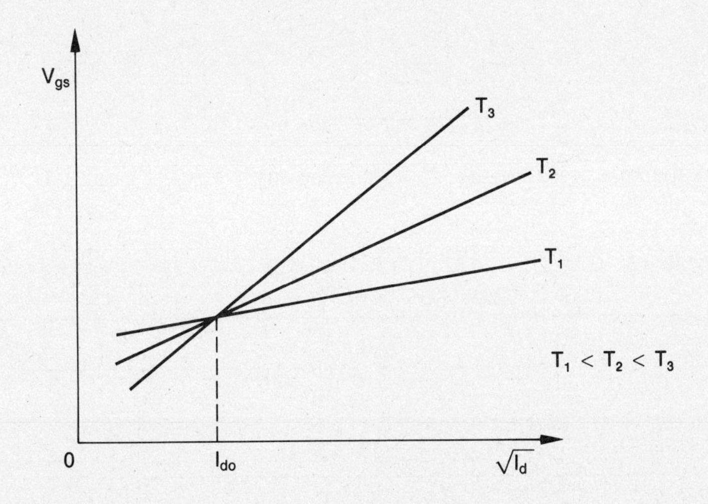

Figure 7.24 Variation of drain current with temperature for enhancement or depletion transistors

Table 7.4 Temperature characteristics of the modified inverter constant voltage reference

Parameter ratio	*Temperature °C*			
	−1	23	53	71
V_{REF}	1.001	1	0.999	0.997
I_d	1.022	1	0.933	0.865

Note:
V_{dd} = 5V
V_{bs} = 0V

An advantage of this simple constant voltage reference is that it can be decentralized, that is, instead of a single reference being mounted throughout a chip, distributed voltage references can be used consisting, in the case of a converter, of several circuits (as shown in Figure 7.25). To obtain higher voltages, the transistor aspect ratios can be changed or enhancement transistors stacked, either within or external to the negative feedback loop to the depletion transistor gate.

Figure 7.26 and Tables 7.5 and 7.6 give results for a nominal 3.5 volt reference used in a converter.

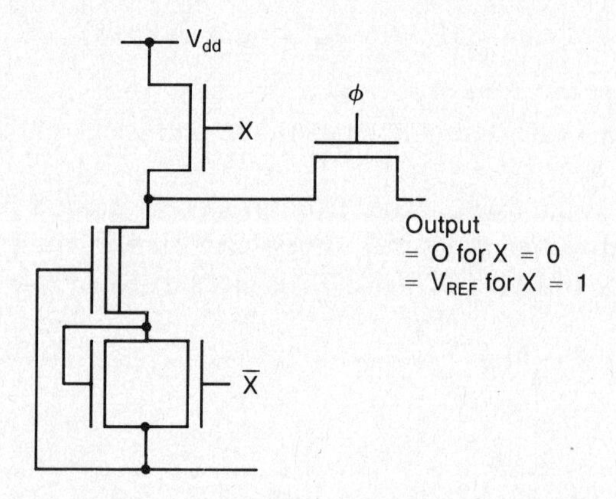

Figure 7.25 Circuit for a distributed constant voltage reference

Table 7.5 Comparison of no load voltage and output resistance for three chips from a single batch

Chip No.	*V_{REF} (no load) volts*	*Output resistance k ohm*
1	3.62	1.1
2	3.65	1.4
3	3.55	1.3

Note:
V_{dd} = 10 volts
T_{amb} = 20.5°C

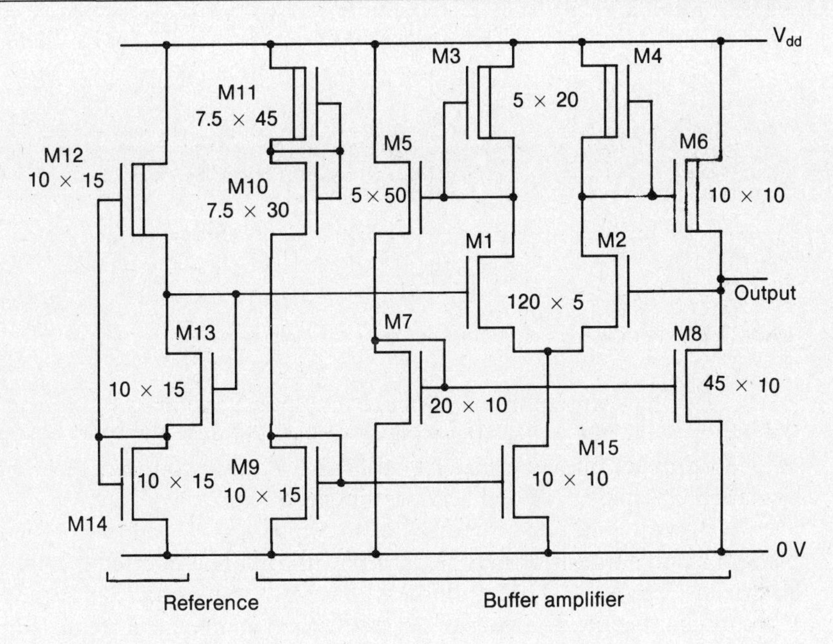

Note:
Dimensions are in microns, with device width given first.

Figure 7.26 A simple 3.5 V voltage reference circuit

Table 7.6 Regulation sensitivity for circuit of Figure 7.26

Input voltage unregulated	5.0	5.5	6.0	7.0	8.0	9.0	10.0	11.0	12.0	13.0	14.0	15.0
Output voltage REF	4.17	3.76	3.63	3.60	3.60	3.61	3.62	3.63	3.64	3.64	3.64	3.65

Note:
$T_{amb} = 22°C$

Turning now to the CMOS process, the normal method of producing a constant reference is to use the band gap reference technique (Vittoz, 1982; Hamilton and Howard, 1975). The voltage across a diode although near constant, falls with increasing temperature. This can be compensated for by adding a series resistor and passing a current through the series diode resistor combination (Figure 7.27) that increases linearity with temperature. The resultant output voltage can be made near constant and approximates, as we will see later, the band gap voltage for silicon (V_g). Although the band gap voltage is nonlinear at very low temperatures, over a normal operating temperature range, we can approximate the band gap voltage for silicon (Hamilton and Howard, 1975) by following linear expression:

$$V_g = 1.205 - 0.28.10^{-3}T \quad \textbf{(7.27)}$$

where T is the temperature in °K.

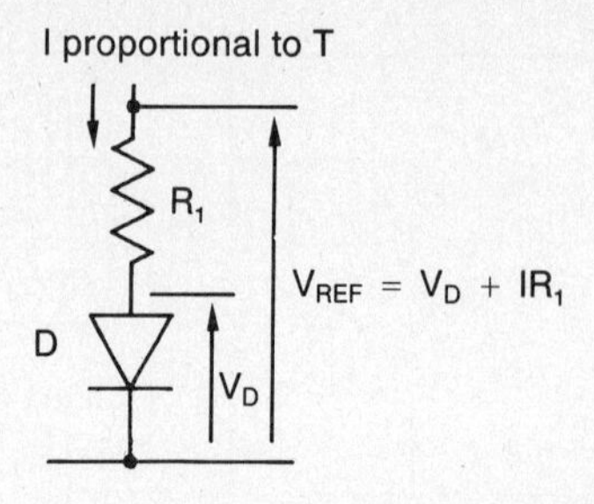

Figure 7.27 Principle of operation of a band gap reference voltage

Two different approaches can be employed: one makes use of an operational amplifier, while the other uses simple single components. We will examine this second approach in some detail to establish the method and then show that the operational amplifier method achieves similar results.

How is a current proportional to temperature produced? It is achieved by using a bipolar transistor current mirror with different emitter current densities (as shown in Figure 7.28). The different current densities can be achieved either by using different emitter areas (parallel transistors) and equal currents (Figure 7.28(a)) or identical transistors and unequal currents (Figure 7.28(b)). In either case, the voltage across the resistor R, being the difference of two base emitter diode potentials, sets the value for current I. Thus,

$$\Delta V_{be} = V_{be1} - V_{be2} = IR_2$$

Using the equation for a diode (3.4), we obtain for either circuit the following:

$$I = \frac{kT}{qR_2}\ln N \tag{7.28}$$

giving a current proportional to temperature, provided R_2 has a zero temperature coefficient.

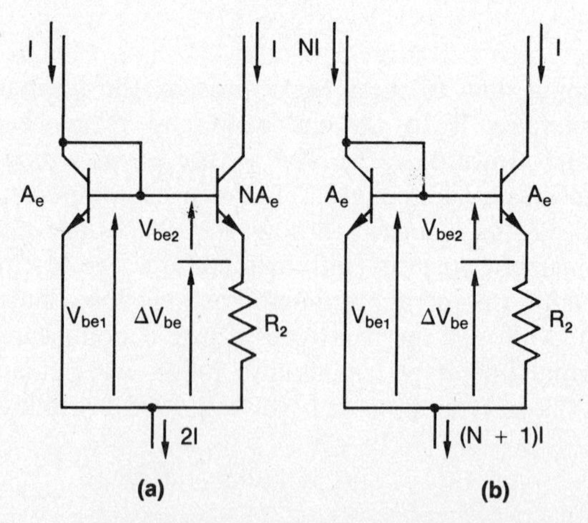

Figure 7.28 Circuits to achieve a current proportional to temperature

Referring back to Figure 7.27, the output voltage is:

$$V_{REF} = V_D + IR_1 \qquad \textbf{(7.29)}$$

Combining with equation 7.28, we obtain the following:

$$V_{REF} = V_D + \frac{R_1}{R_2} \times \frac{k\ln(N)T}{q} \qquad \textbf{(7.30)}$$

Thus, the output voltage is the diode voltage plus a voltage proportional to temperature, where the proportionality constant is fixed by the ratio of two resistors and two emitter current densities, that is, geometry considerations.

Using again the diode equation for V_D (equation 3.4) and an expression for the diode leakage current (Meijer, 1982), we obtain the following:

$$I_0 = KT^n e^{-qV_g/kT} \qquad \textbf{(7.31)}$$

where K is a proportionality constant and n is typically 1.4. We can expand equation 7.30 to become,

$$V_{REF} = V_g + \frac{kT}{q}\ln\left(\frac{I}{KT^n}\right) + \frac{R_1}{R_2}\left(\frac{k\ln(N)T}{q}\right) \qquad \textbf{(7.32)}$$

We require V_{REF} to have a zero temperature coefficient at some ambient temperature T_0. This is achieved by differentiating equation 7.32 with respect to temperature and then back substituting the resultant conditions, giving,

$$V_{REF0} = 1.205 + (n - 1)kT_0/q \qquad \textbf{(7.33)}$$

While this is the voltage at temperature T_0, we wish to know the change in reference voltage as the temperature changes from T_0. Again this is achieved by differentiating equation 7.32, giving,

$$\frac{\delta V_{REF0}}{\delta T} = \frac{k}{q}(1 - n) \times \left(1 - \frac{T_0}{T}\right) \qquad \textbf{(7.34)}$$

By plotting these results (Figure 7.29), we can see that the reference voltage has a very small temperature coefficient.

An alternative form of this circuit, developed by National Semiconductor (National Semiconductor, 1971; Soclof, 1985) is shown in Figure 7.30. Transistors Q_1 and Q_2 with R_3 generate the current I, while the base emitter diode of Q_3 acts as the series diode in Figure 7.27. Assuming the collector and emitter currents of Q_2 are equal, we have the following:

$$V_{REF} = V_{be} + \frac{R_2}{R_3}\Delta V_{be}$$

$$= V_{be} + \frac{R_2}{R_3} \times \frac{k\ln(N)T}{q}$$

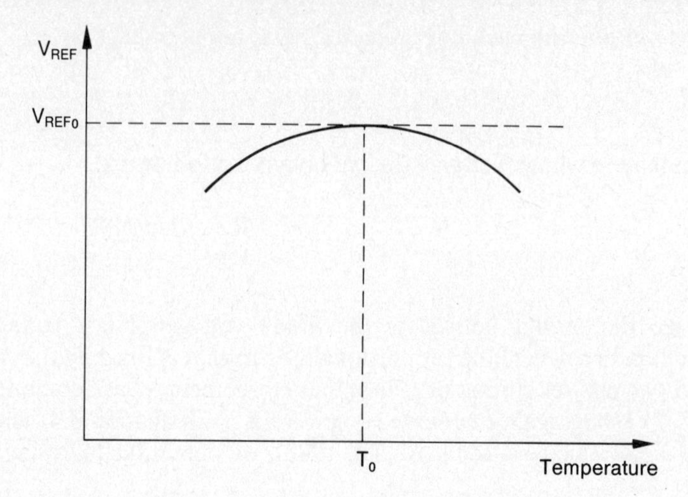

Figure 7.29 Output voltage for a band gap reference voltage circuit

Comparing with equation 7.30 of the previous circuit, we can see that they are of identical form and therefore this second circuit must provide a similar function, namely a constant output voltage. Other circuit variations are possible.

The second approach to a band gap reference is to use an operational amplifier. Figure 7.31 gives two examples of this approach, one employing two diodes the other two transistors. The figure gives the equations for the output voltages and again they are of the same form as equation 7.31. There are, however, two noticeable differences. The first, a disadvantage, is that the derivation assumes no offset voltage for the amplifier. Should this not be the case, then the equations have added to them an offset term. The second difference is an advantage, in that the amplifier gain can be used to scale the output voltage. This has been done for reference (b), using resistors R_5 and R_6, but could also be done for reference (a).

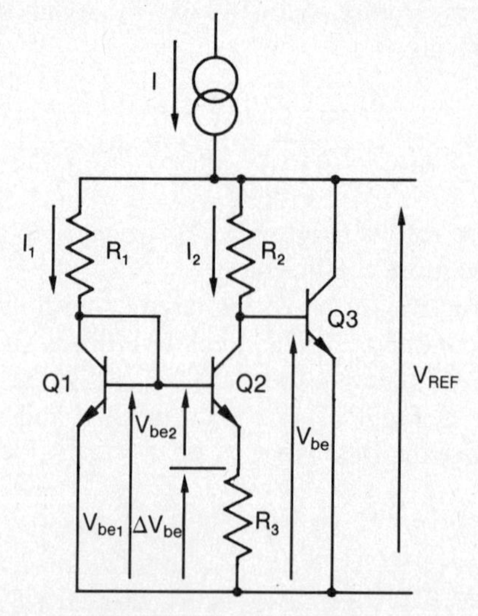

Figure 7.30 Alternative band gap reference circuits

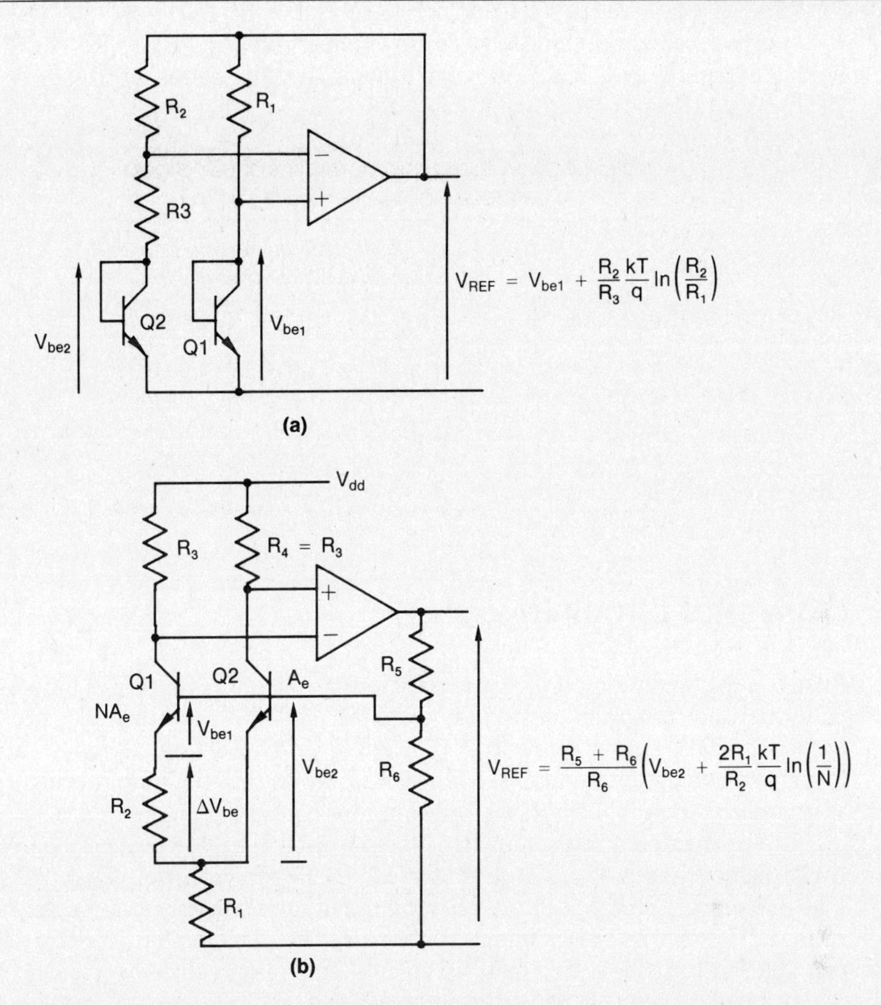

Figure 7.31 Band gap constant voltage references employing operational amplifiers

Tables 7.7 and 7.8 provide results for two band gap circuits. Table 7.7 is for the circuit given in Figure 7.30, with resistors made from p^+ diffusion. Here, R_1 = 1220 ohm, R_2 = 6500 ohm and R_3 = 440 ohm. MOS type lateral transistors are used.

The circuit is driven by a MOS current mirror, giving I = 1 mA.

Table 7.7 Measured characteristics for a nonoperational amplifier band gap reference (Figure 7.30)

Temperature °C	−2°	25°	35°	50°
Regulated output volts	1.42	1.42	1.41	1.39

Note:
I_{drain} = 1 mA

The second circuit is given in Figure 7.31(a). Here R_1 = 2632 ohm, R_2 = 7896 ohm and R_3 = 564 ohm, giving resistor ratios of $R_2:R_1$ = 3 and $R_2:R_3$ = 14. Results are given in Table 7.8.

Table 7.8 Measured characteristics of band gap reference circuit of Figure 7.31(a)

Unregulated input volts	*Regulated output (volts) at various ambient temperatures*		
	5°C	25°C	50°C
3.0	1.205	1.225	1.274
4.0	1.211	1.229	1.276
5.0	1.213	1.230	1.276
6.0	1.215	1.232	1.276
7.0	1.218	1.232	1.276

7.6 Converter circuit layouts

Plate 5 is the layout of a three-bit resistor string converter, which was drawn using a parameterized program, where the variable "nBITS" controls the size of the converter. The output provided is inverted, with 000 digital input producing maximum analog output and 111 zero volts output, thus the output being $-V_{STRING}$ as is required for a following switched capacitor converter (Figure 7.8).

The measured maximum error for a five-bit converter using this program is −0.75 percent when V_{REF} = 2 volts, and −0.90 percent when V_{REF} = 2.5 volts. The maximum error occurs at maximum output voltage and is mainly due to voltage drops across the switching transistors. The magnitude of the error suggests that the resistor string is well matched to a tolerance better than 1 percent.

Plate 6 shows the layout of a three-bit capacitor analog to digital converter. A dummy capacitor is included to feed in the input from a resistor string converter (see Figures 7.6 and 7.8).

7.7 Exercises

1 The serial charge redistribution digital to analog converter is difficult to implement in silicon because of stray capacities to substrate, associated with the sources and drains of the switches. Derive an expression for the output voltage with the stray capacities added and thus compute the errors they introduce.

2 For the transistor current mirror converter of Figure 7.11, derive an expression for the output voltage V_0.

3 The resistor string analog to digital converter (Figure 7.19) employs an encoder to give the digital output. Derive the logic for such an encoder. Compare your solution with the alternative of simply feeding the shift register clock in parallel into a binary counter, to generate the digital output.

4 Refer to Figure 7.7. Draw the layout for a capacitor with associated switches that could be used as the basis for constructing the weighted capacitor structure of this converter.

5 Refer to Figure 7.32(a). The input capacitor C is associated with a parasitic capacitor C_p. Using simple analysis, compute the output voltage for the two cases given below when C_p is placed in alternative positions.

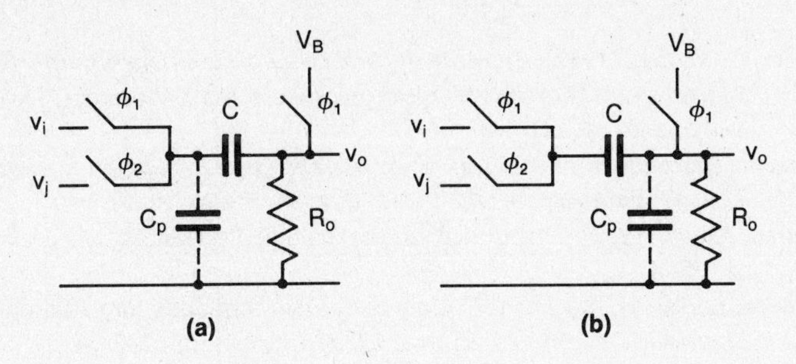

Figure 7.32

6 For the voltage reference circuit of Figure 7.22, derive the full expression for the output voltage V and so determine the proportionality constant in equation 7.22.

7 Two simple voltage reference circuits are given in Figure 7.21 and 7.23. For the case of identical transistor geometry and where $W = 5\ \mu m$ and $L = 10\ \mu m$, complete a SPICE analysis of the circuit, comparing how the output voltage changes with: (a) input voltage; (b) load; and (c) temperature.

8 Derive the output expressions for the two band gap reference circuits of Figure 7.31.

7.8 References

Albarran, J. F. and Hodges, D. A. (1976), "A Charge Transfer Multiplying Digital to Analogue Converter", *IEEE Journal of Solid State Circuits*, Vol. SC–11, pp. 772–9.

Allstot, D. J. (1982), "A Precision Variable-Supply CMOS Comparator", *IEEE Journal of Solid State Circuits*, Vol. SC–17, pp. 1080–7.

Blanschild, R. A., Tucci, P. A., and Muller, R. S. (1978), "A New nMOS Temperature Stable Voltage Reference", *IEEE Journal of Solid State Circuits*, Vol. SC–13, pp. 767–74.

Fotouhi, B. and Hodges, D. A. (1979), "High Resolution A/D Conversion in MOS/LSI", *IEEE Journal of Solid State Circuits*, Vol. SC–14, pp. 920–6.

Gregorian, R. (1981), "High Resolution Switched Capacitor D/A Converter", *Microelectronics Journal*, Vol. 12, No. 2, pp. 10–13.

Hamilton, D. J. and Howard, W. G. (1975), *Basic Integrated Circuits*, McGraw-Hill, New York.

Haskard, M. R. (1983), "A Simple nMOS Constant Voltage and Current Source", *Microelectronics Journal*, Vol. 14, No. 4, pp. 31–7.

Hoff, M. E., Higgins, J. and Warren, B. M. (1979), "An nMOS Telephone Codec for Transmission and Switching Applications", *IEEE Journal of Solid State Circuits*, Vol. SC–14, pp. 47–53.

Huang, J. Y. (1980), "Resistor Termination in D/A and A/D Converters", *IEEE Journal of Solid State Circuits*, Vol. SC–15, No. 6, pp. 1084–7.

Jain, L. C. and Haskard, M. R. (1984), "Review of D/A and A/D Converters and Their Suitability for VLSI", *Journal of Electrical and Electronic Engineering (Australia)*, Vol. 4, No. 1, pp. 35–42.

McCreary, J. L. and Gray, P. R. (1975), "All MOS Charge Redistribution Analogue to Digital Conversion Techniques—Part 1", *IEEE Journal of Solid State Circuits*, Vol. SC–10, pp. 371–9.

Meijer, G. C. M. (1982), *Integrated Circuits and Components, for Bandgap References and Temperature Transducers*, Electronics Research Laboratory, Delft University, Delft, The Netherlands, March.

National Semiconductor (1971), *IC Provides On-Card Regulation for Logic Circuits*, Application Note 42, February.

Rehman, M. A. (1980), "Integrated Circuit Voltage References", *Electronic Engineering*, Vol. 52, pp. 65–85.

Sherman, L. (1984), "12 bit A/D Converter Slips Smoothly into Analogue and Digital Worlds", *Electronic Design*, Vol. 32, No. 22, 31 October, pp. 227–38.

Soclof, S. (1985), *Analogue Integrated Circuits*, Prentice-Hall, Englewood Cliffs, NJ.

Suarex, R. E., Gray, P. R. and Hodges, D. A. (1975), "All MOS charge Redistribution Analogue to Digitial Conversion Techniques—Part 2", *IEEE Journal of Solid State Circuits*, Vol. SC–10, pp. 379–85.

Vittoz, E. A. (1982), "The Design of High Performance Analogue Circuits on Digital CMOS Chips", *IEEE Journal of Solid State Circuits*, Vol. SC–17, pp. 1080–7.

Yee, Y. S., Terman, L. M. and Heller, L. G. (1978), "A ImV MOS Comparator", *IEEE Journal of Solid State Circuits*, Vol. SC–13, pp. 294–8.

Yukawa, A. (1985), "A CMOS Eight Bit High Speed A/D 779 Convertor IC", *IEEE Journal of Solid State Circuits*, Vol. SC–20, p. 775.

8 Oscillators and phase locked loops

8.1 Introduction

Almost every chip requires an oscillator even if it is simply for the generation of a two phase nonoverlapping clock. In some cases, however, there may already be a signal present and the chip must either synchronize with this signal or operate at a subharmonic of the frequency available. With the strong emphasis today on digital circuits, usually it is not necessary to generate sine wave oscillators as square wave types are adequate. In this chapter, we will consider oscillator circuits, whose frequency of oscillation can be controlled by a voltage, threshold circuits and simple phase locked loops.

8.2 Oscillator circuits

8.2.1 Oscillator types

Two classes of oscillator circuits will be considered. In most designs, the frequency of oscillation either needs to be precisely set to a single frequency or can be varied over a wide range, using an externally or internally generated control signal. Thus the classes of oscillators to be considered are, first, a range of voltage controlled oscillators (VCO) and, second, crystal controlled oscillators, where the frequency of oscillation is accurately set by an external piezo electric crystal.

8.2.2 Ring oscillators

The simplest type of oscillator is an extension of the ring oscillator, found on the starting frame of most multiproject chips, used to determine the constant τ and hence calculate delays in digital circuits. It consists of an odd number (typically 19) of inverters (nMOS or CMOS) connected in a ring, oscillating at a frequency of around 10 MHz, depending on the particular MOS process parameters. By building in series and/or shunt resistor capacitor networks it is possible to lower the frequency and, thus, make it voltage controllable. Figure 8.1 shows two possible circuits. With circuit (b), the control voltage can be taken down to near the enhancement transistor's threshold voltage so that the resistance becomes very large. Consequently, the range of frequency control can be extremely large. For an M stage ring oscillator, the delay per stage is,

$$t = RC\ln\left(1 - \frac{V_{th}}{V_{dd}}\right) \tag{8.1}$$

and hence the frequency of oscillation is,

$$f = \frac{1}{2MCR\ln\left(1 - \frac{V_{th}}{V_{dd}}\right)} \tag{8.2}$$

where V_{th} is the threshold voltage of an inverter. Figure 8.2 illustrates this point, showing the range of frequencies available from a five stage type (b) ring oscillator,

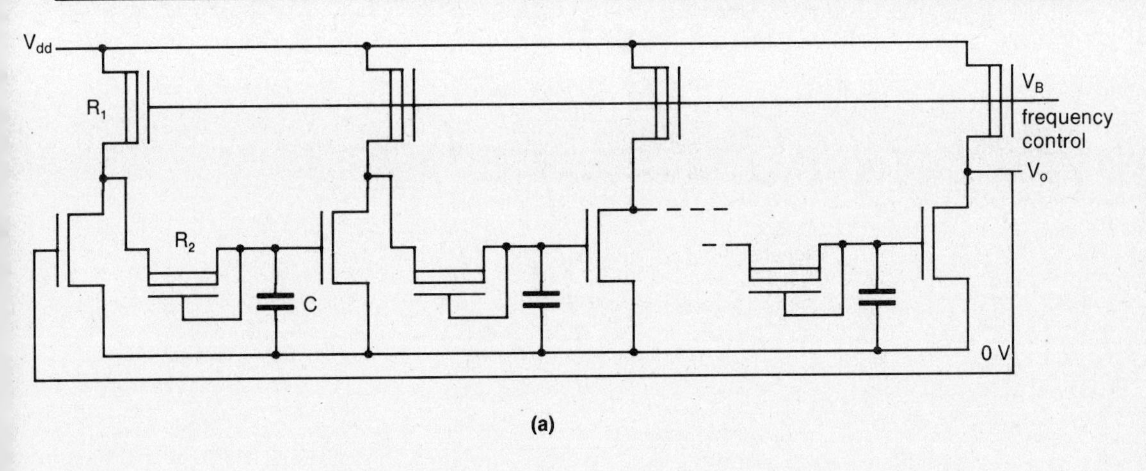

(a)

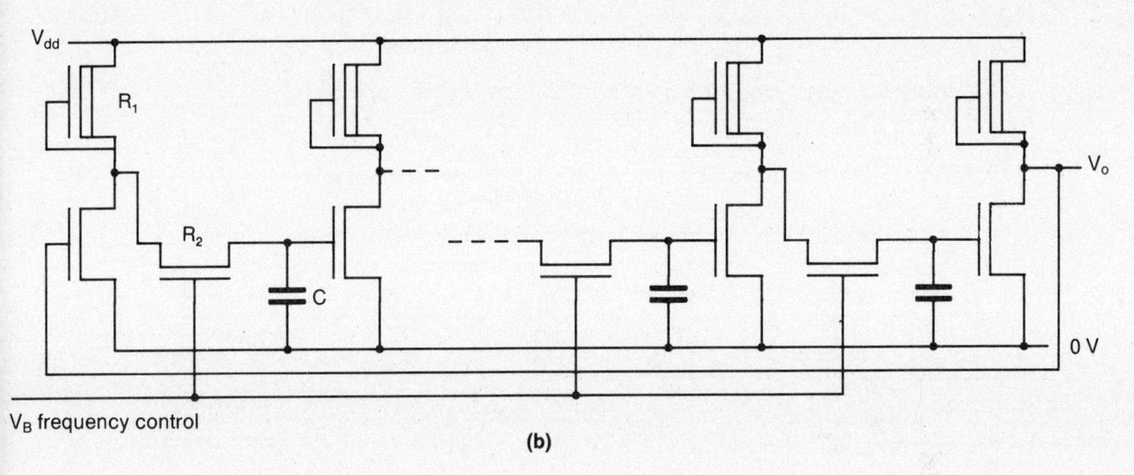

(b)

Note:
For both circuits, the number of stages must be odd.

Figure 8.1 nMOS ring oscillators

using standard nMOS tiles. The minimum frequency of oscillation is normally set by the loop gain falling below unity, often being lower than the minimum self start up frequency of oscillation. The graph given in Figure 8.2 is reproduced only for the range where self starting occurs.

This particular circuit is very useful for generating a two phase nonoverlapping clock. Consider the case of seven inverters in a ring. Each of the seven outputs is delayed in time with respect to each other. Consequently, by combining outputs (as shown in Figure 8.3), an on chip clock generator can be made with a single external pad required to control the frequency. A second pad may be employed to bring one phase of the clock out for observation or synchronization of other circuits.

The circuit given in Figure 8.1(a) offers a narrower range of control for the same input control voltage swing, which can be advantageous.

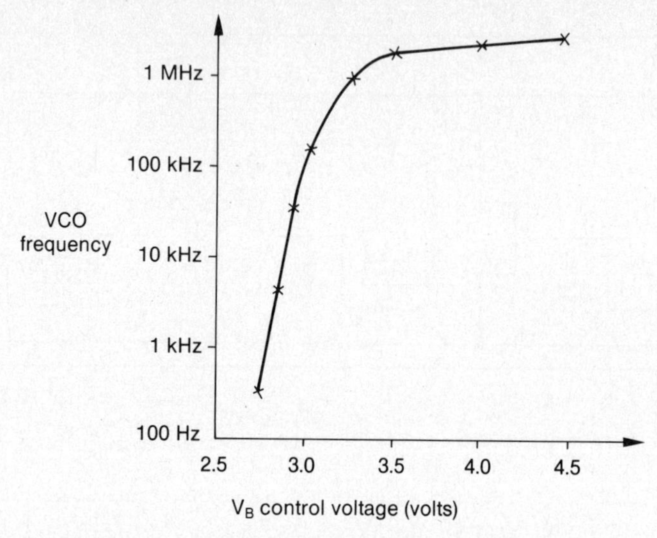

Figure 8.2 Frequency range of a VCO ring oscillator of five stages

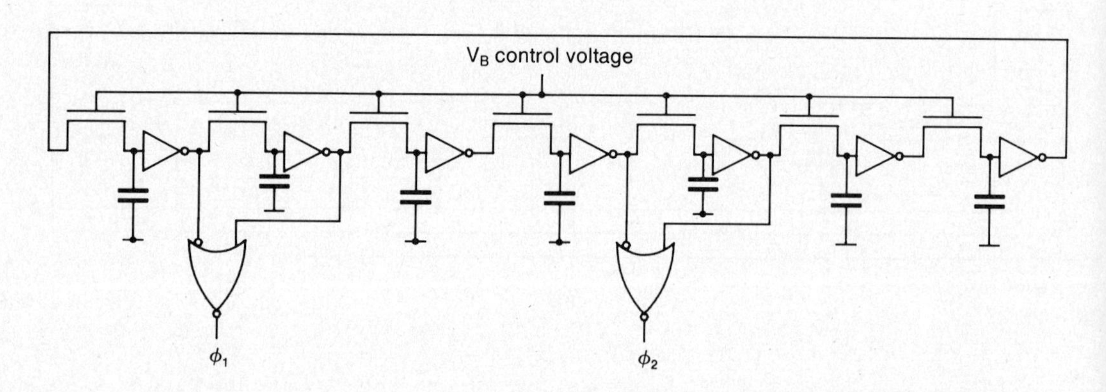

Figure 8.3 An on chip two phase clock generator using a voltage controlled ring oscillator

8.2.3 Constant current charging oscillators

An alternative way of producing a voltage controlled oscillator is to charge or discharge a capacitor with a voltage controlled current. To do this, the modified inverter constant voltage reference circuit is used as a constant current generator circuit (as shown in Figure 8.4). Initially, capacitor C is discharged with the reset precharge circuit. The lower plate of the capacitor is now at V_{dd} and acts as a supply rail for the constant current generator circuit. As a result, the potential on the lower capacitor plate falls toward zero and at some point the threshold circuit is triggered, resetting the circuit. Several threshold circuits may be used to provide outputs at different points in time thereby also allowing the generation of a two phase nonoverlapping clock. To put delay into the reset circuit, either a bistable or simple RC delay monostable circuit must be used (as illustrated in Figure 8.5).

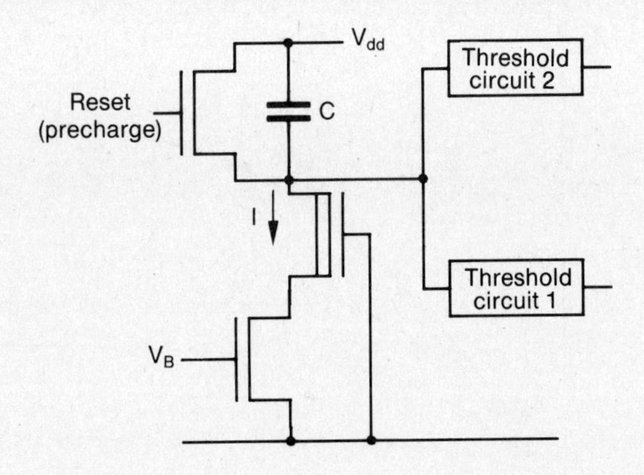

Figure 8.4 Simple constant current discharge circuit which can be used as the basis of an oscillator

This particular circuit has the advantage of good linearity between input control voltage and source current and hence output frequency. This is illustrated in Figure 8.6.

The remaining portion of the circuit yet to be discussed is the threshold circuit. In principle, inverters of different aspect ratios (Figure 8.7(a)), may be used. However, the modified inverter circuit (Figure 8.7(b)) provides an alternative solution if only two threshold levels are required, as it is not difficult to set the depletion transistor gate voltage to either zero or V_{dd}. For a 9:1 aspect ratio, threshold voltages of approximately 1 and $2\frac{1}{2}$ volts result. Tiles are provided for these circuits. A third alternative is the regenerative comparator (bistable) approach, where the threshold voltage is again determined by the inverter aspect ratios that make up the bistable.

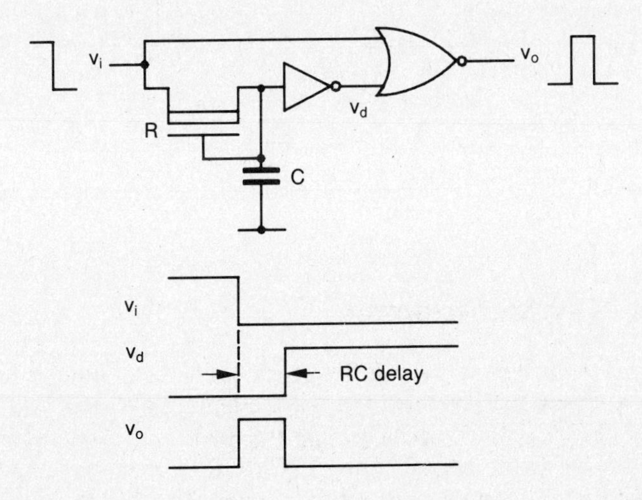

Figure 8.5 Simple RC pulse generating reset circuit

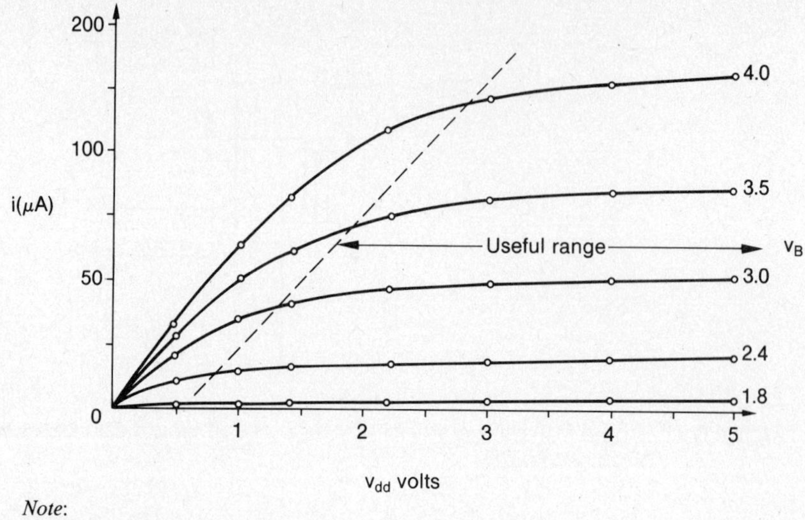

Note:
Both devices W/L = 2.

Figure 8.6 Relationship between source current and input control voltage for the modified inverter current source

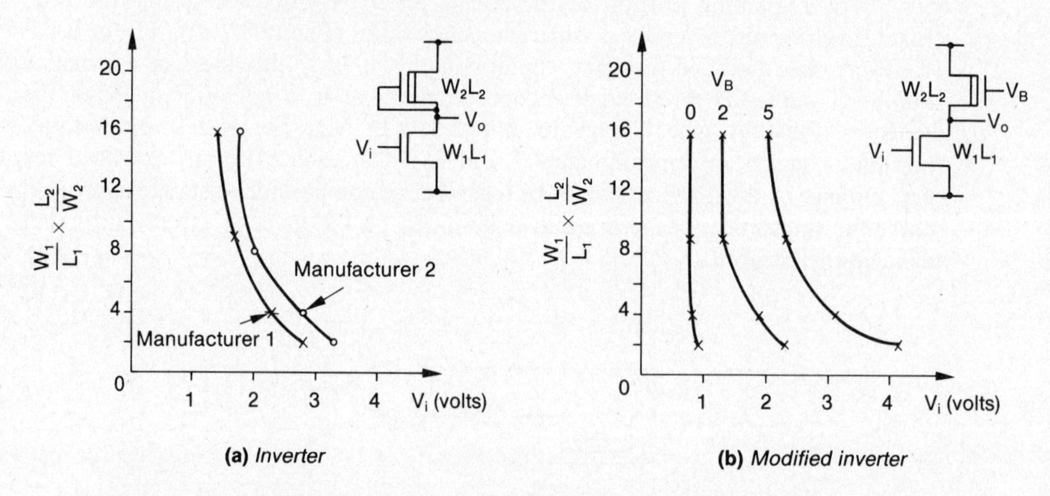

Figure 8.7 Threshold voltage of an inverted and a modified inverter circuit

8.2.4 Crystal oscillators

Many systems require a single frequency oscillator of a stable and known frequency. Here a crystal oscillator can be used. Figure 8.8 shows two circuits (Santos and Meyer, 1984) both of which require two pads and two external capacitors (C_1 and C_2), as well as the crystal. For the CMOS process, the large resistance depletion transistors can be made enhancement types or a resistor can be made from the high resistance well material.

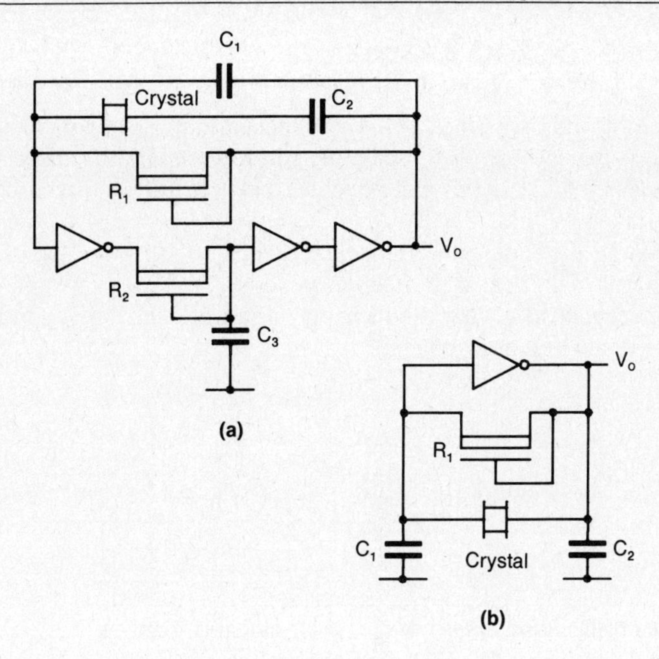

Figure 8.8 Crystal oscillator circuits using standard inverters

Considering in more detail the Pierce oscillator circuit (Mattys, 1983) given in Figure 8.8(b), the depletion transistor resistor should have a value greater than 100 kilo-ohm and typically 1 Mega-ohm. For nonsinusoidal oscillation, C_1 can be omitted and C_2 fixed at 15 pF. For devices made on the nMOS Manufacturer 2 process, crystal oscillators of up to at least 4 MHz are possible. For a 1 MHz, sine wave oscillator $C_1 = 8$ pF, $C_2 = 15$ pF and $R_1 = 1$ Mega-ohm. Santos (1984) provides a detailed analysis of this circuit.

Two phase nonoverlapping clocks can be generated from these circuits using either a four-bit twisted ring counter or a three-bit self correcting shift register and appropriate decoding logic. The shift register version is shown in Figure 8.9. Standard six gate D type flip flops must be employed. Note that the two phase clock frequency is one quarter that of the input oscillator frequency.

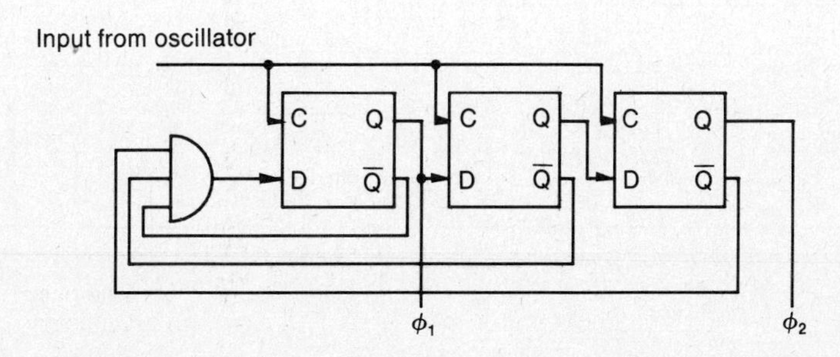

Figure 8.9 Circuit to generate a two phase nonoverlapping clock supply from an oscillator

8.3 Phase locked loops

The basic building block of a phase locked loop is shown in Figure 8.10. It consists of a voltage controlled oscillator, phase comparator/mixer and low pass filter. Frequency dividers (N) can be added between the voltage controlled oscillator and the mixer so that it will run at a multiple ($N \times f_{in}$) of the input frequency f_{in}. Similarly, by adding the frequency divider in the input leg, the voltage controlled oscillator will run at a frequency some multiple lower (f_{in}/N) than the input frequency. In this way, frequency synthesizers and other similar units can also be incorporated on a silicon chip.

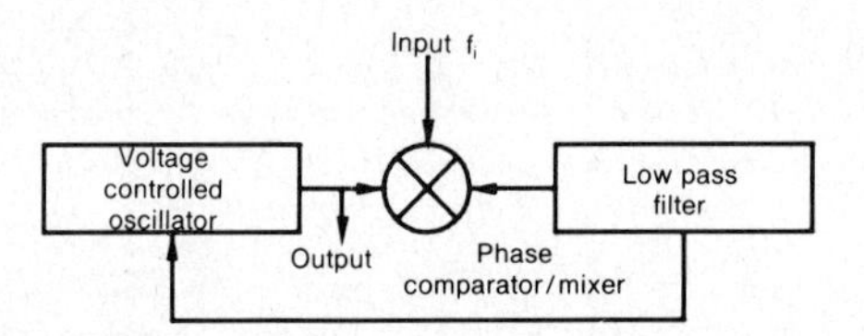

Figure 8.10 Basic elements of a phase locked loop

Returning to the basic phase locked loop of Figure 8.10, because non-sinusoidal oscillators are used, the mixer can be a simple gate. Further, as already seen, the control voltage input line of the voltage controlled oscillator is high impedence and capacitive so that it can be fed directly from a passive low pass filter. Figure 8.11 gives a simple circuit.

In this case, the phase comparator is an "OR" gate which has a response, as shown in Figure 8.12. When lock occurs at the center of the range, there will be a 90° phase shift between the input frequency and the voltage controlled oscillator output. Other comparator types, such as an exclusive OR gate or master slave JK flip flop (Rohde, 1983) can be employed. Note that the oscillator circuit has two input lines, the additional one being used to trim the center frequency.

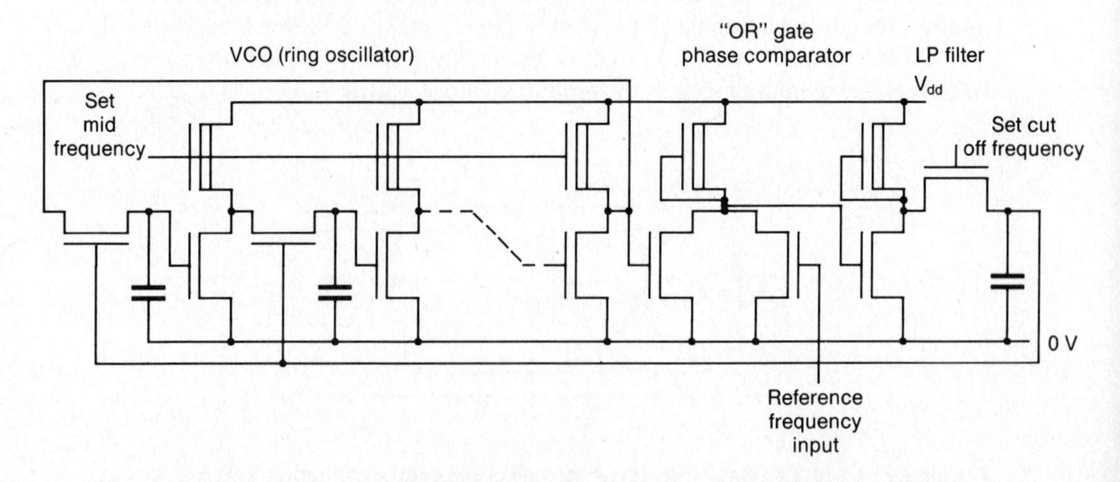

Figure 8.11 Simple phase locked loop

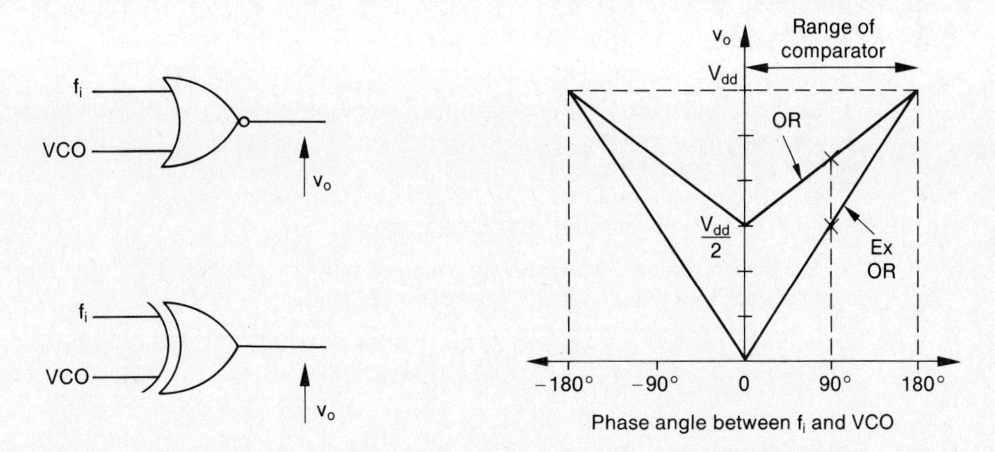

Figure 8.12 Operation of simple OR and exclusive OR gate phase comparator

While a simple analog approach has been used for the filter, it is possible to use digital techniques such as a set/reset bistable and up/down counters.

8.4 An oscillator circuit constructed from tiles

Results for various circuits employing selected tiles from the appendix have been already given. In general, all of the circuits discussed can be assembled from the tiles. As an example, Figure 8.13 shows the layout of one stage of the five stage ring oscillator, of which the frequency characteristics are reported in Figure 8.2.

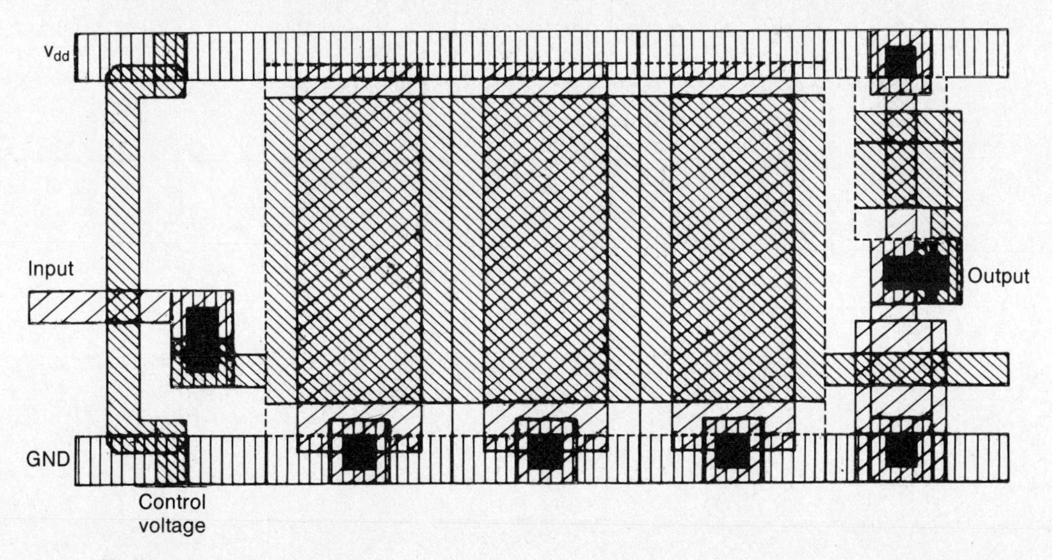

Note:
All stages interconnect by abutment.

Figure 8.13 One stage of a ring oscillator showing the voltage controlled MOS resistor, gate oxide capacitor and 9:1 inverter

8.5 Exercises

1 Refer to Figure 8.4. Calculate the frequency of the voltage controlled oscillator, assuming the following:
(a) a threshold voltage V_t;
(b) the constant current $I = KV_B$ where K is a proportionality constant; and
(c) the reset monostable introduces a delay t_d.

2 Using the method given in Figure 8.4, design the necessary reset logic, employing a flip flop, to make a voltage controlled oscillator.

3 Produce the circuit for a frequency synthesizer which will give out any frequency in the range of 10 kHz to 100 kHz, in 10 kHz steps.

8.6 References

Matthys, R. J. (1983), *Crystal Oscillator Circuits*, Wiley-InterScience, New York.
Rohde, U. L. (1983), *Digital Phase Locked Loop Frequency Synthesizers: Theory and Design*, Prentice-Hall, Englewood Cliffs, NJ.
Santos, J. T. and Meyer, R. G. (1984), "A One Pin Crystal Oscillator for VLSI Circuits", *IEEE Journal of Solid State Circuits*, Vol. SC–19. No. 2, pp. 228–36.

9 Sensors

9.1 Why the need for sensors

A large number of systems being designed today are for instrumentation and control applications. These systems must be capable of monitoring the processes they control and to do so, they require sensors. For example, take the automobile. It is relatively easy to install a microprocessor chip to control the engine and vehicle performance, provided the engine and vehicle parameters can be measured. Thus, sensors are required for pressure, temperature, speed or revolutions per second, braking force, light level, acceleration, etc. While sensors already exist for these various parameters, they are usually expensive, nonlinear and are affected by ambient conditions. It is indeed fortunate that silicon has characteristics that make it suitable for many sensor applications (Regtien, 1978). Since the same material is used for making microelectronics circuits, it is possible to integrate the sensor with appropriate electronics to reduce or eliminate the sensor's deficiencies and perform appropriate computations on the data measured. Smart sensors are bringing with them a new era in instrumentation.

With most MOS sensor designs, the designer does not have the freedom to change the fabrication process parameters to optimize the sensor but must instead use a standard MOS process and design the sensor around it. What then are the properties of sensors designed in this way? Many fabricators are unable or unwilling to provide the designer with the information required. Consequently, the sensor performance can only be crudely estimated and confirmed later by actual fabrication. Despite reported work on what can be expected from standard multiproject chip nMOS processes (Haskard, 1983), the problem of interfacing of sensors to electronic circuitry still remains. Further, other sensor types can be made which do not depend on the properties of the silicon parent material. These include the metalization, capacitive and resistive types, which may be used for measuring acceleration, pressure force, displacement, conductivity, dew point, and so forth (Haskard, 1984; Kaiser and Proebster, 1908). Some systems use sensors in arrays rather than having a single device. Examples of this are a photo diode array, used as an electronic eye for a robot, or a Hall effect array, used to read a magnetic card. In this chapter, we will examine sensors that depend on silicon properties, sensors that do not, and finally the problem of interfacing of electronic circuitry to the sensors.

9.2 Silicon sensor types

The most important properties of silicon are the temperature, photoelectric, piezo resistive and Hall effects. We will consider sensors based on these.

9.2.1 Thermal sensors

The simplest temperature dependent components are the diffusion and polysilicon resistors. Here, the temperature dependent resistance is represented by (Hamilton and Howard, 1975):

$$R(T) = \frac{KLT^{\eta}}{qnWt} \tag{9.1}$$

where K = a proportionality constant
L = the resistor length
W = the resistor width
t = the resistor thickness
q = the electron charge
n = the carrier density
T = the temperature in degrees Kelvin
η = the power constant

Figure 9.1 shows the temperature characteristics of resistors from two different fabricators. Comparing these results with equation 9.1 and noting that the reference indicates that η is also temperature dependent, measured values for η over the temperature range 0 to 50°C vary from 0.45 to 0.60 for n^+ diffusion and 0.17 to 0.24 for polysilicon resistors. Inspection of the results suggests that a satisfactory approximation for modeling characteristics of the resistors could be obtained by simply applying linear regression techniques. Thus, for polysilicon resistors, the slope is typically 7.58×10^{-4} and intercept is 0.775, while for n^+ diffusion resistors, the slope is 1.80×10^{-3} and intercept is typically 0.470.

An alternative approach when making a thermal sensor is to use either the reverse or forward characteristic of a diode. With the nMOS process, it is unusual to use a forward biased diode. However, measurements made on diodes with a forward bias current of $10\ A/m^2$ gave a forward voltage temperature coefficient of -2.26 mV/°C.

Considering now a reverse biased diode, since the leakage current depends heavily on minority carriers, one would expect considerable variations from one manufacturer to another. This is the case (refer to Figure 9.2) for there is a

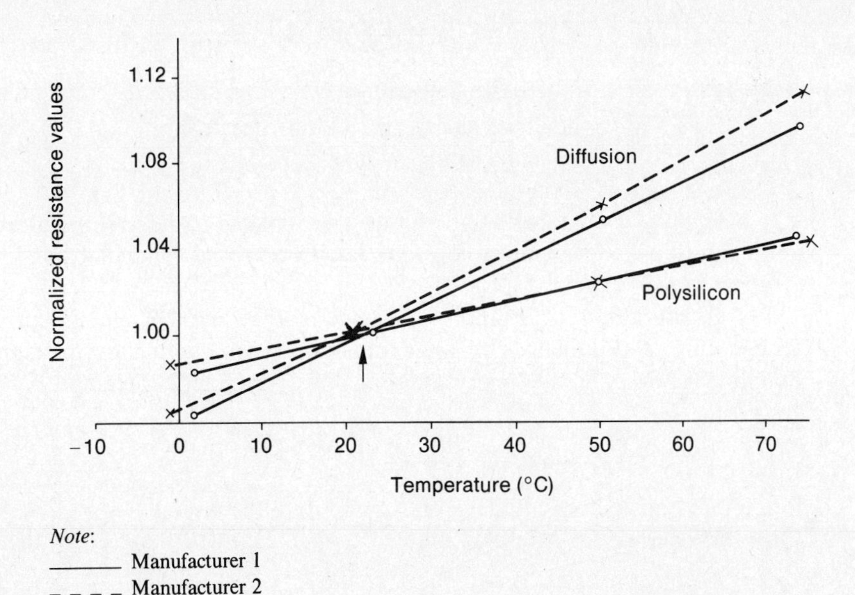

Figure 9.1 Temperature characteristics of n^+ diffusion and polysilicon resistors of an nMOS process

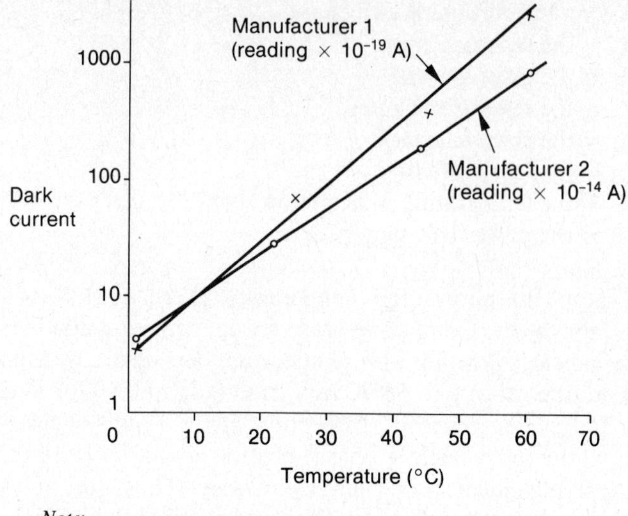

Note:
Diode dark current per square micron

Figure 9.2 Reverse bias diode leakage currents as a function of temperature

measured difference of five orders of magnitude between the two nMOS manufacturer's samples.

An expression often used to model a diode leakage current (Meijer, 1982) is:

$$I_0 = KLWT\gamma e^{-qV_{go}/kT} \tag{9.2}$$

where K = a proportionality constant
L and W = the diode length and width respectively
T = the temperature in degrees Kelvin
k = Boltzman's constant
V_{go} = the band gap voltage extrapolated to zero temperature
γ = the power constant, having a typical value of 1.5

Examination of Figure 9.2 suggests that the change in current due to temperature is dominated by an exponential term, with currents approximately doubling for every 8°C temperature rise. Thus, a simpler expression of,

$$I_0 = C_1 LWe^{-C_2 qV_{go}/kT} \tag{9.3}$$

is a more satisfactory model here. For Manufacturer 1, $C_1 = 2.37 \times 10^8$ and $C_2 = 0.876$ while for Manufacturer 2, $C_1 = 1.21 \times 10^{11}$ and $C_2 = 0.772$.

The third thermal sensor possibility is the MOS transistor. The rate of change of the gate source voltage with respect to temperature is given by (Blauchild et al., 1978):

$$\frac{dV_{gs}}{dT} = -\alpha + \sqrt{\frac{L}{kC_{ox}W}}\frac{d}{dT}\sqrt{\frac{I_d}{\mu}\left(\frac{T}{T_0}\right)^{\frac{3}{2}}} \tag{9.4}$$

where α = the temperature coefficient of the threshold voltage
μ = the carrier mobility

Thus, below a given drain current level, the temperature coefficient of the gate source voltage is negative, while above it, the temperature coefficient is positive. Measurements made on both enhancement and depletion transistors of drawn width 10 μm and length 5 μm indicate that the zero temperature coefficient occurs at very low currents, typically 0.3 μA to 2.5 μA. The temperature coefficient for the enhancement transistor in the positive region is typically 12.5 mV/°C.

9.2.2 Optical sensors

Junction diodes can be operated in a photoconductive or photovoltaic mode. The light generated current is given by (Green, 1982):

$$I_L = qLWG(L_e + L_h + \omega) \tag{9.5}$$

while the open circuit voltage is,

$$V_L = \frac{kT}{q}\ln\left(\frac{I_L}{I_0} + 1\right) \tag{9.6}$$

where G = the generation rate of electron hole pairs
L_e and L_h = the diffusion length of electrons and holes
ω = the width of the depletion region

Two diode types have been examined, one with overglaze and the other without (Figure 9.3(a)). Measurements indicate that the overglaze had little effect on the sensitivities and spectral response (Figure 9.3(b) and (c)). However, it does appear that the thickness of the second "oxide" layer between the polysilicon and metal does play an important part. This is surprising since capacity measurements suggest both oxides are of similar thickness. Table 9.1 compares typical photovoltaic characteristics of 100 μm square diodes.

Table 9.1 Sensitivities of 100 μm square photo diodes to a tungsten filament lamp

			Manufacturer 2		
	Manufacturer 1		Run 5/82		Run 5/83
Parameter	Overglaze	No overglaze	Overglaze	No overglaze	Overglaze
Open circuit volts (V)	0.464	0.463	0.120	0.123	0.07
Short circuit current (μA)	1.60	1.25	0.75	0.80	0.95

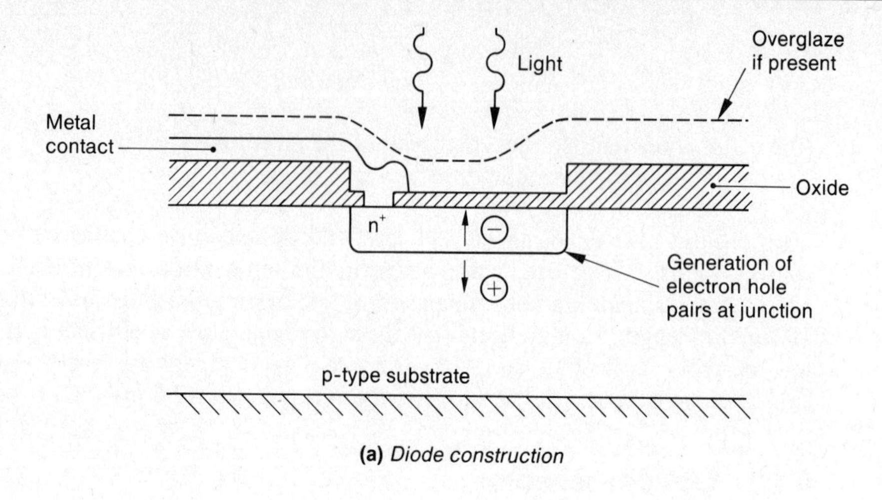

(a) *Diode construction*

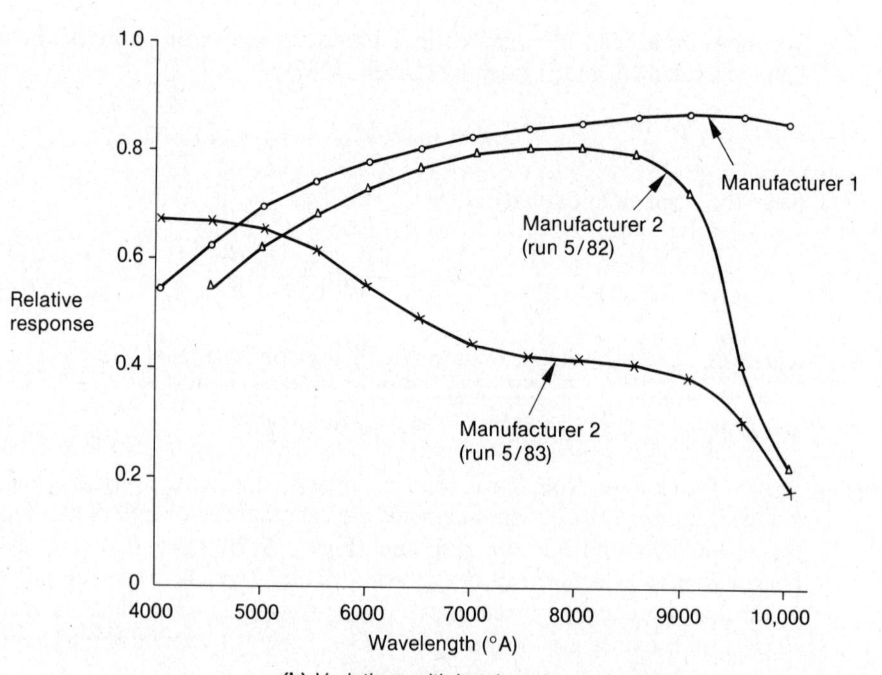

(b) *Variations with batch and manufacturer*

(*continued*)

9.2.3 Magnetic sensors

The conventional method of designing Hall effect devices is to use the high resistivity epitaxial layer of a bipolar process to achieve a large mobility factor. Unfortunately, this layer is not available on the nMOS process, but the channel region of a transistor can be used instead. The experimental depletion transistor device is shown in Figure 9.4. It is unconventional in that it can be operated in either a voltage or current mode.

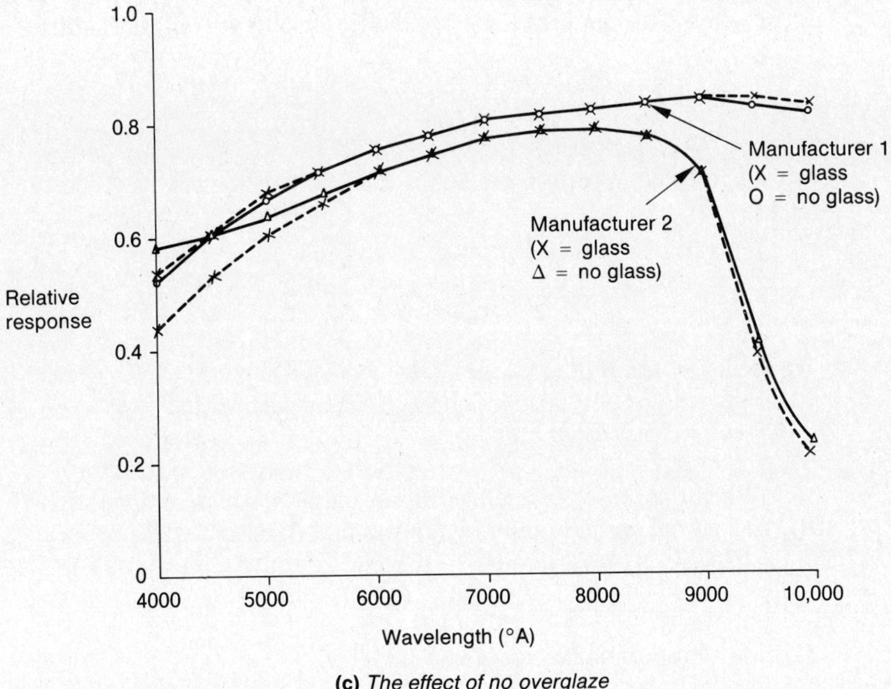

(c) *The effect of no overglaze*

Notes:
Frequency response = 100 μm square
Photo diodes = 100 μm square
Temperature = 20°C

Figure 9.3 Construction and spectral response of $N^{+}p$ photo diodes

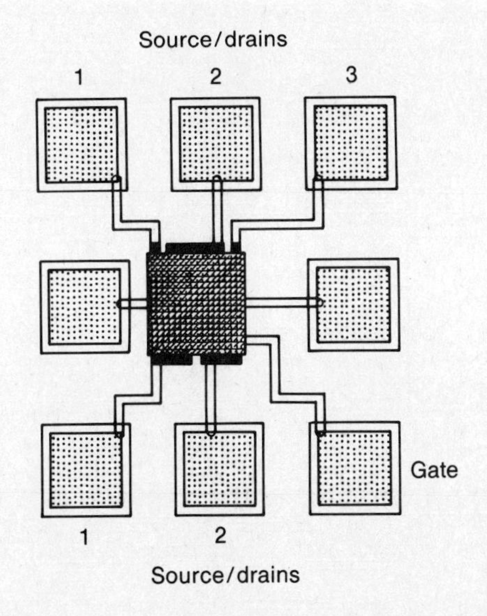

Figure 9.4 Experimental Hall effect depletion transistor sensor

In the current driven voltage mode, the output Hall effect voltage is given by,

$$V_H = \frac{R_H IB}{D} \tag{9.7}$$

while in the voltage driven mode, the same voltage is given by (Chien and Westgate, 1980),

$$V_H = \frac{V_s \mu BW}{L} \tag{9.8}$$

where R_H = the Hall coefficient
B = the normal component of the magnetic field
μ = the carrier mobility.

When the sensor is operated in an output current configuration (Figure 9.5), the Hall effect current is given by (Davies and Barnicoat, 1970),

$$I_H = I_1 - I_2 = C\mu BI \tag{9.9}$$

C being a proportionality constant.

Figures 9.6 and 9.8 show typical characteristics of sensors operated in all three modes. For MOS smart sensors, the current modes appear to be the preferred method.

The offset voltage and current arise from two causes: first, the errors in the photolithographic process resulting in a lack of symmetry and, second, those due to the piezo resistive properties of the silicon. When the sensor is mounted, using epoxy material, stresses can be induced in the silicon during curing, causing the sensor to be out of balance. To cancel out this, offset subsidiary gate regions can be added (Braun et al., 1974).

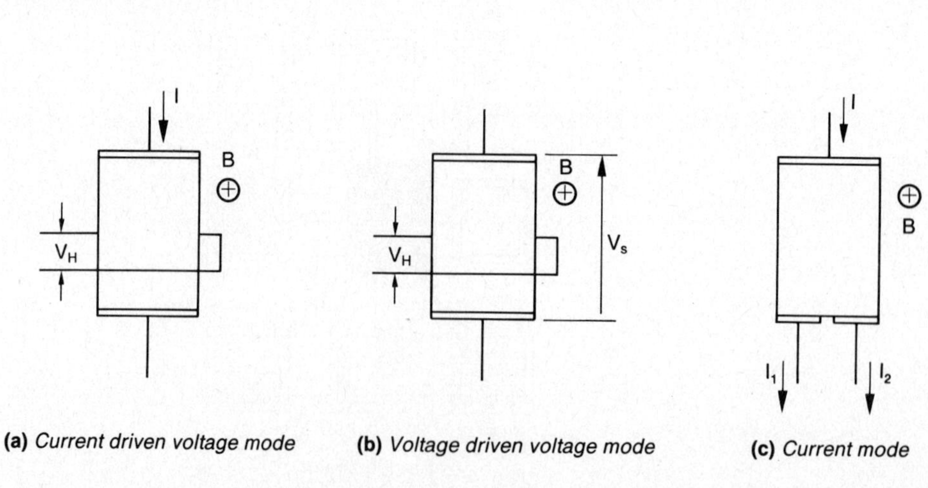

Figure 9.5 Methods of operating Hall effect sensors

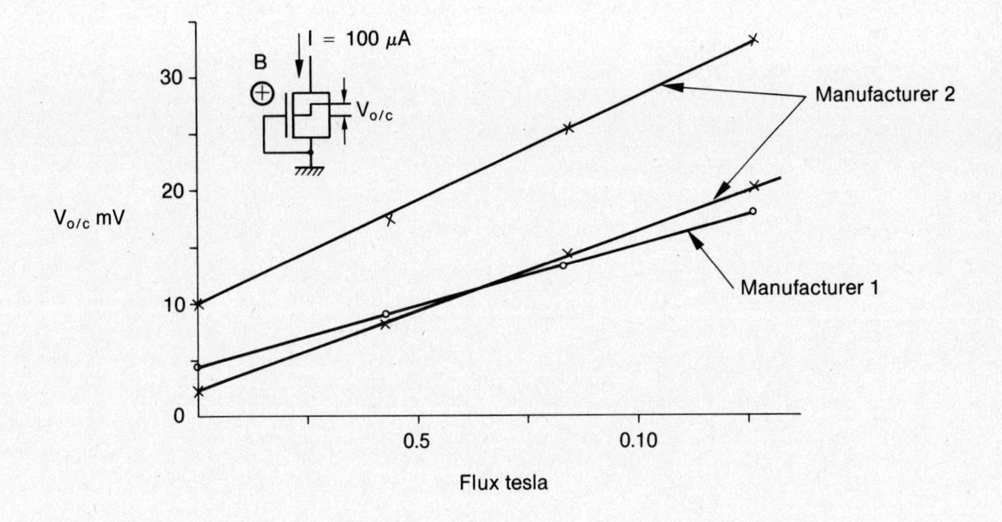

Note:
Bias current = 100 μA.

Figure 9.6 Spreads in Hall effect output voltage both within a batch and from manufacturer to manufacturer

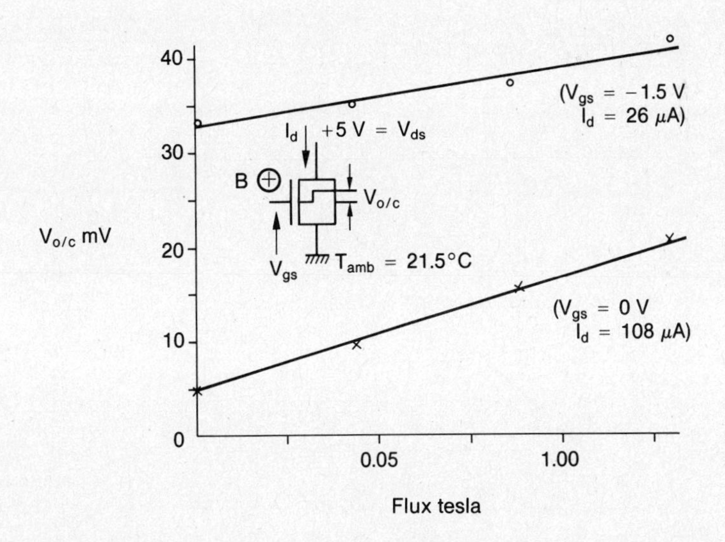

Note:
Drain source bias = 5 volts.

Figure 9.7 Variations in Hall output voltage with gate bias

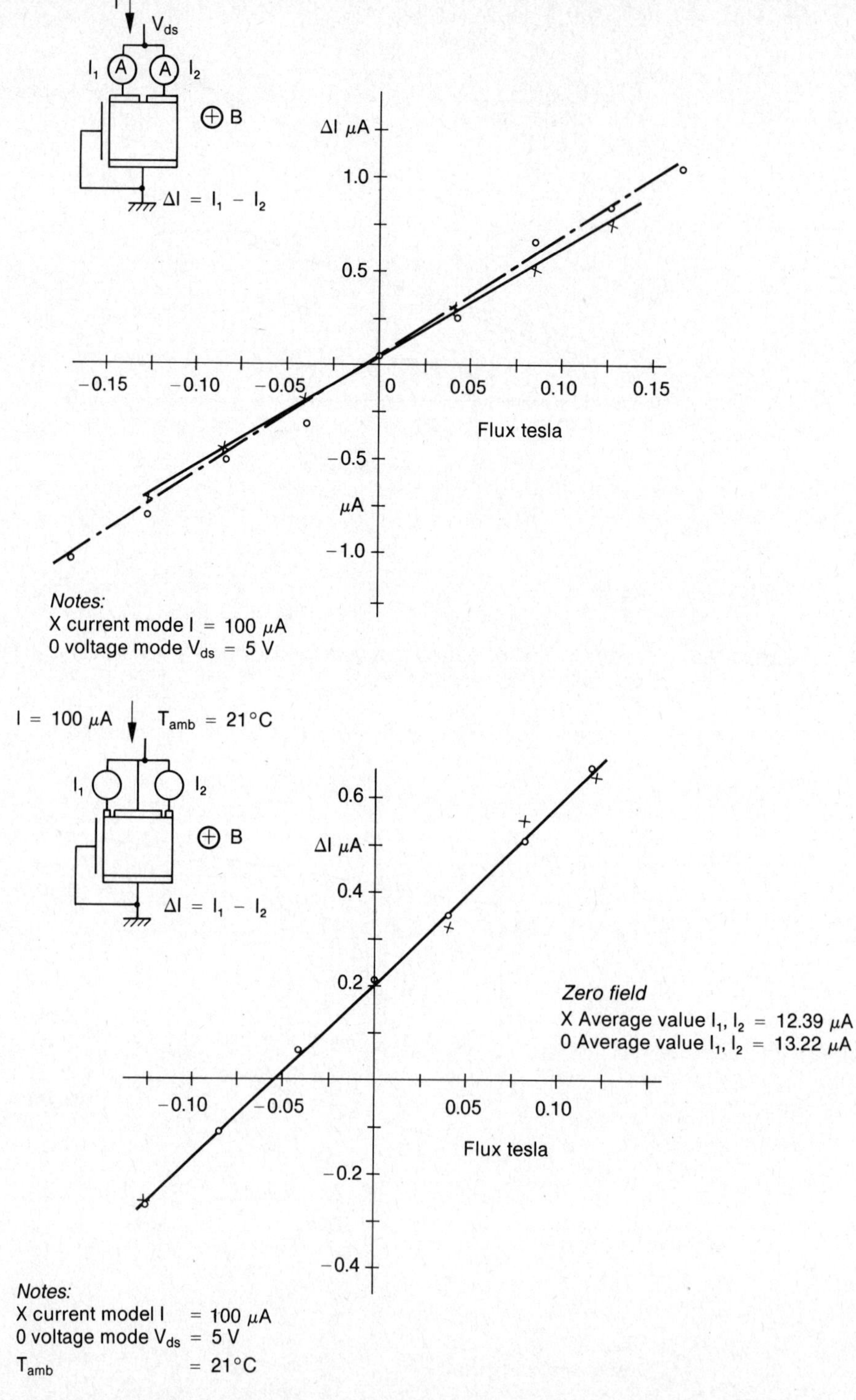

Figure 9.8 Hall effect current for two different manufacturers and two different geometry ratios

9.2.4 Strain sensors

The piezo resistive properties of silicon can be exploited by using diffusion, polysilicon and depletion transistors. The gauge factor G for a semiconductor piezo resistance device is given by (Window and Holister, 1982),

$$G = \frac{\Delta R / R}{\epsilon} \tag{9.10}$$

where ΔR = the change in resistance
R = the material resistance
ϵ = the axial strain applied

Figures 9.9. and 9.10 show the results of strain measurements for devices fabricated by two foundries. Even though measurements were repeated, all showed nonlinear characteristics at initial low strain levels, suggesting that the adhesion material may not be transmitting the correct stress. With depletion transistors, this effect was not evident, even though the same proprietory epoxy strain gauge adhesive was used.

There are difficulties in mounting some types of nMOS sensors, the strain sensor being a good example. The normal gauge adhesive is nonconductive and, yet, electrical contact needs to be made to the substrate to set the substrate voltage. The difficulty could be overcome if a substrate connection is made to the top surface of the chip. It would probably be a high resistance contact, but this would not matter as current flow is negligible. For multiproject chips, it is suggested that a substrate contact be placed at the end of the starting frame.

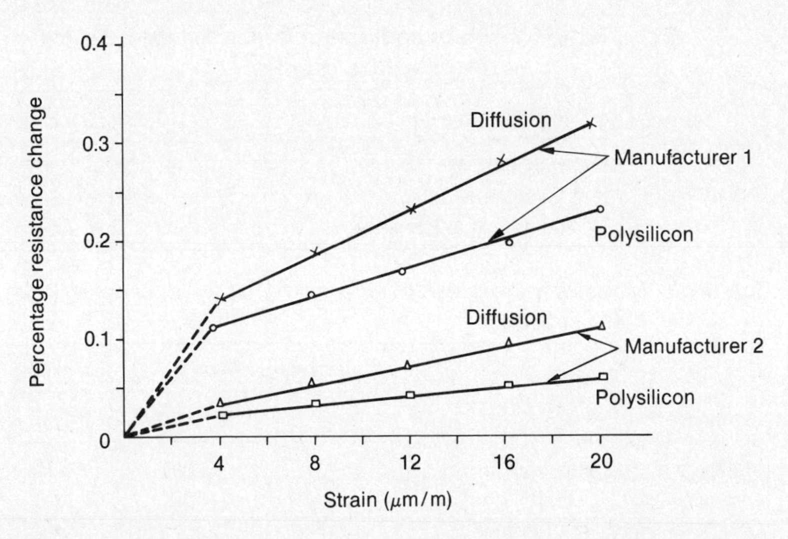

Note:
Change in resistance with tension strain, T_{amb} = 18 °C

Figure 9.9 Resistance change N^+ diffusion and polysilicon strain gauges

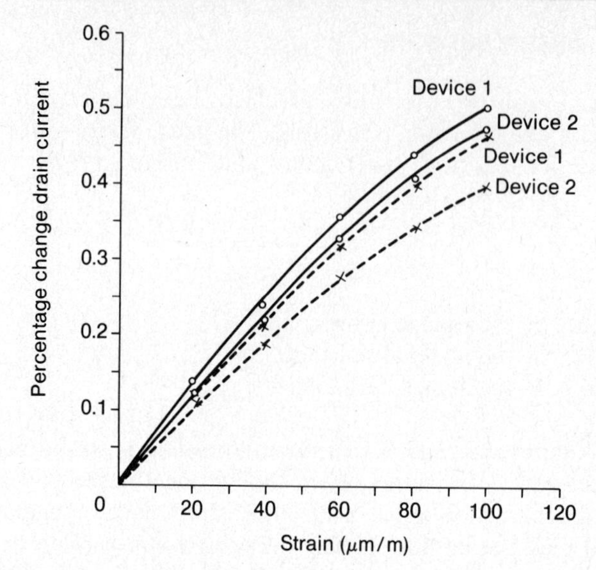

Notes:

1. Depletion field effect transistors, matched pair, manufacturer 1:
   V_{ds} = 5V
   V_{sb} = 0
   T_{amb} = 20°C
2. _ _ _ _ $V_{gs} = 0$
   ——— $V_{gs} = 5V$

Figure 9.10 Drain current charges for *n*MOS long depletion transistor strain gauges

Let us return to the experimental sensors. The measured ranges of gauge factors are shown in Table 9.2, while Table 9.3 provides matching information on long depletion transistor devices.

Table 9.2 Measured gauge factors for various nMOS process strain gauges

Strain gauge device	*Gauge factor range*
n^+ diffused resistor	−10 to −70
Polysilicon resistor	−10 to −60
Depletion nMOS transistor	−20 to −40

Table 9.3 Measured piezo resistance parameters for long depletion mode nMOS transistors

	Manufacturer 1		*Manufacturer 2*	
Parameter	$V_{gs} = 0$	$V_{gs} = 5V$	$V_{gs} = 0$	$V_{gs} = 5V$
Matching initial static drain circuit	0.6%	2.0%	42%	—
Average sensitivity × 10^{-3} μm/m	2.9	3.1	1.8	3.8
Percentage matching of sensitivity	15	7.5	52	—

Notes:
V_{ds} = 5V
V_{sb} = 0V
T_{amb} = 20°C

9.3 Nonsilicon sensor types

In some cases sensors can be made using nonsilicon components of a chip. The most important of these is the upper metalization layer, used to either form one plate of a capacitor, or metal interdigitized fingers, to measure resistance of fluids or other conducting medium placed between the fingers. The capacitance transducer is very powerful and a chip with a capacitance plate can be used as part of a sensor to measure many parameters, such as pressure, acceleration and displacement. (See Figures 9.11 and 9.12.)

Associated with the metalization plate on the chip is a parasitic capacity to substrate or even to the polysilicon layer. Increasing the area of the capacitor sensor also increases the associated parasitic capacitor so the ratio of maximum sensor capacity to parasitic capacity remains constant, the value depending on the dielectric thicknesses and constant.

The capacitor sensor is also useful as there are many ways it can be integrated into a system. Circuit examples include bridges, oscillators, charge/discharge circuits and switched capacitor networks. Some of these methods will be discussed in more detail in section 9.4.

A very useful property of a FET is the extremely high gate input resistance. Consequently, floating gate circuits can be used to measure static electric fields or they can be interfaced through a gold electrode into physiological systems to monitor nerve and other body potentials. Alternatively, the gate can be coated with a carbon polymer thick film paste to form an ion selective electrode (Hoffman, et al., 1984). The paste can be sensitized to respond to the desired ion. In the abovementioned illustrations, the gold and carbon filled polymer paste can be placed

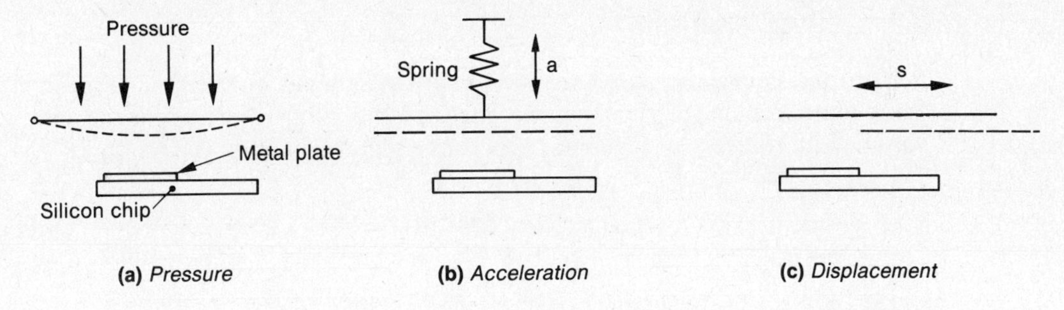

Figure 9.11 Capacitance sensors used to determine pressure, acceleration and displacement

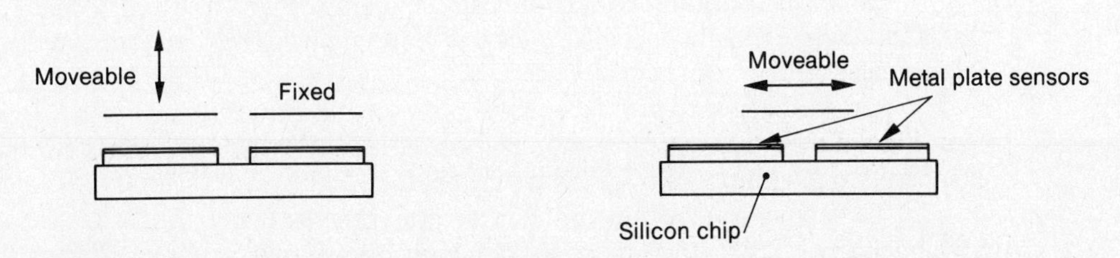

Figure 9.12 To possible forms of a differential sensor

directly onto the chip or on a separate substrate which is electrically connected. When the former is used, care must be taken to suitably encapsulate the chip so that it remains insulated from the environment.

9.4 Interfacing to sensors

Since most sensors are analog in nature, signals derived from them must be converted into a digital form. It is assumed in this section that the chip in which the sensor is incorporated or interfaced to has some intelligence, and, therefore, no matter what format the digital output from the sensor may be in, the digital portion can interpret it and apply correction factors to improve linearity. Corrections for environmental effects such as temperature may also be applied, but it is often better to employ a differential sensor system to cancel out such effects.

To convert the sensor output into a digital signal, a number of methods are possible. These include:

1. conventional analog to digital conversion;
2. a ring oscillator;
3. voltage (or current) to frequency (or time) conversion;
4. lateral inhibition.

As conventional analog to digital conversion methods have already been discussed in Chapter 7, this section will deal with the three remaining methods only.

9.4.1 Ring oscillators

Perhaps the simplest system is to use a ring oscillator (as discussed in Chapter 6) where one or more of the frequency determining components are made into a sensor. With reference to Figure 9.13(a), the period of oscillation is given by,

$$T = N(F_1 + F_2)RC = \frac{RC}{K} \tag{9.11}$$

where N = the number of stages in the loop

F_1 and F_2 = factors depending on the inverter threshold characteristics

R and C = the passive components, one or more of which will be a sensor.

Good temperature correction can be obtained by using a second fixed frequency ring oscillator in a phase locked loop, applying the correction voltage to both oscillators.

If the capacitor to ground is made a sensor, the parasitic capacitor is shunted out. Should the series resistor component be selected as the sensor, it may be made a temperature or strain sensitive resistive element.

To temperature compensate, a second ring oscillator, with fixed passive components, is used in a phase locked loop, the output being the difference in oscillator frequencies. (See Figure 9.14.)

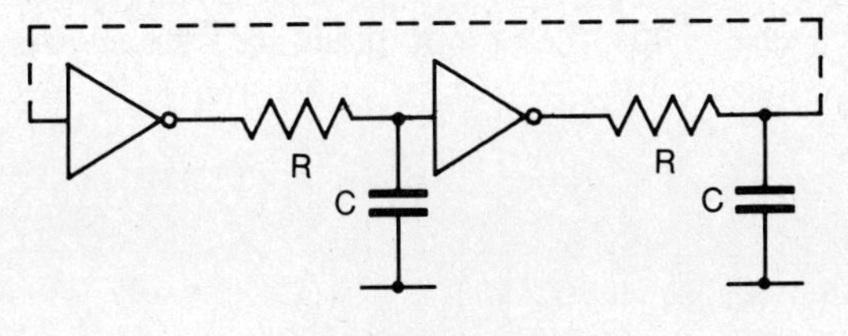

(a) *Ring oscillator*

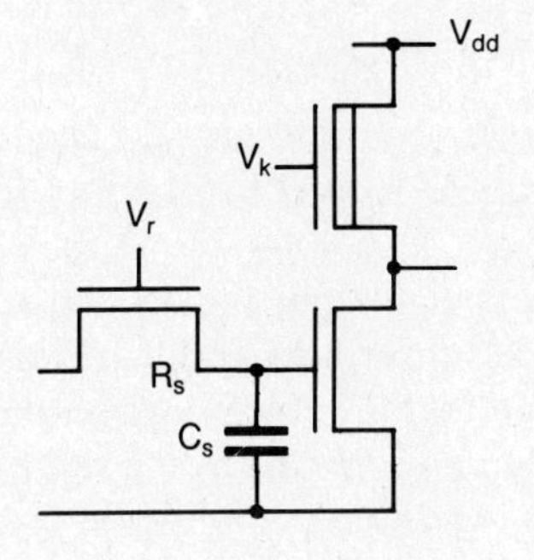

(b) *Detailed single section of a ring oscillator*

Note:
The number of stages N must be odd (only two are shown).

Figure 9.13 Ring oscillator where the frequency of oscillation is varied by the sensors

In the case of a capacitor sensor, the stages of the ring counter can be as shown in Figure 9.13(b). The sensor ring oscillator frequency is,

$$f_s = \frac{K_a}{R_s C_s} \qquad \textbf{(9.12)}$$

where $C_s = C_0 \pm \Delta C$, ΔC being the change in sensor capacity. The dummy oscillator frequency is,

$$f_d = \frac{K_d}{R_d C_d} = \frac{K_s}{R_s C_0} \qquad \textbf{(9.13)}$$

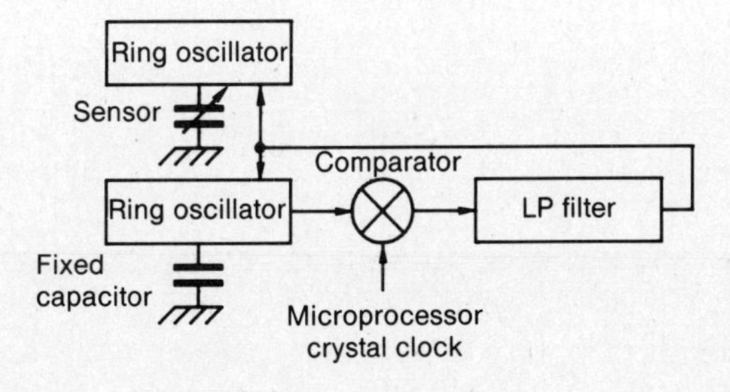

Figure 9.14 Ring oscillator sensor system

for the ideal case. Thus, the output difference frequency is,

$$f = \frac{K_s}{R_s}\left(\frac{1}{C_0} - \frac{1}{C_0 \pm \Delta C}\right)$$

For C_0, much greater than ΔC

$$f = \frac{K_s \Delta C}{R_s C_0} \qquad \textbf{(9.14)}$$

By controlling V_k (and hence K_s), the two oscillators are set to the same initial frequency, while V_r is used to trim any frequency change due to temperature (Figure 9.14).

Ring oscillators can be made into two wire sensors if one remotely monitors the current flow in the supply lines and from this deduce the frequency of oscillation. Again, to compensate for unwanted environmental effects, a second matching ring oscillator is employed and imposed to the same conditions as the first. As discussed, the required signal is the difference in oscillator frequencies and this can be extracted in hardware by employing up/down counters or, in software, by using a microprocessor.

9.4.2 Frequency or time conversion

The reverse biased diode and current mode Hall effect device are two examples of sensors that provide an output current. By integrating these currents with time, a current to frequency or time conversion circuit results. Two examples are given in Figures 9.15 and 9.16.

In the case of Figure 9.15, the frequency of oscillation, ignoring the narrow reset pulse, is,

$$f = I/C_t V_{th} \qquad \textbf{(9.15)}$$

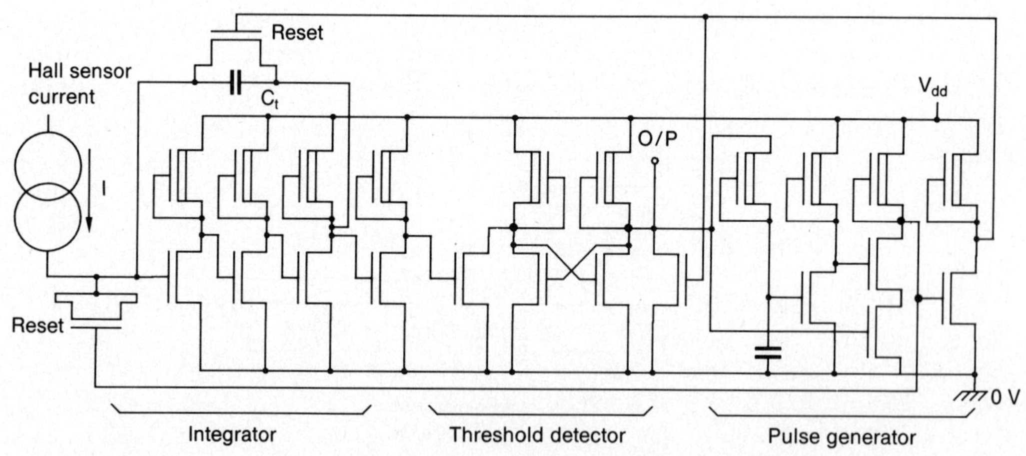

Figure 9.15 Hall effect sensor employing current to frequency conversion

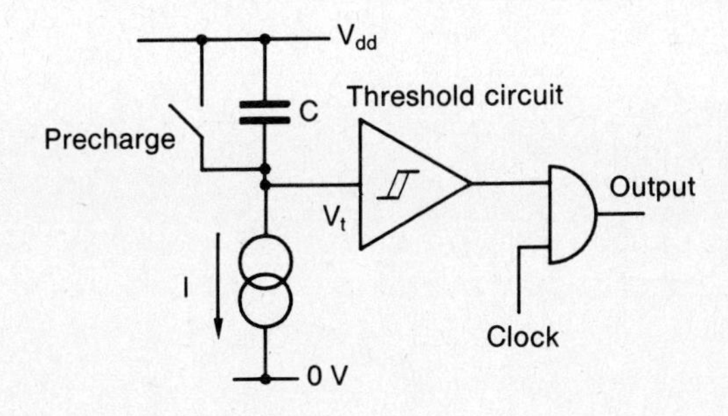

Figure 9.16 Time delay sensor interface circuit

where V_{th} is the threshold voltage of the threshold detector. If the pulse generator is removed, and therefore the reset loop broken, the unit becomes a current to time converter.

Figure 9.16 is a modified voltage controlled oscillator circuit (Figure 8.4), where the rate of discharge of the capacitor is determined either by the sensor current (Regtien, 1978) or, alternatively, the capacitor value, if the capacitor is the sensor.

On precharge, the voltage on the lower plate is charged to V_{dd}. The threshold circuit is set, the "and" gate opened and the clock pulses fed through for counting. The voltage into the threshold circuit is,

$$V_t = V_{dd} - It/C \tag{9.16}$$

At some point, the threshold circuit switches and the clock count out is proportional to time t.

For improved performance, a differential system should be used. In the case of the Hall effect current mode device, two currents are available from the one sensor, which can be directly employed to make a differential system (Cooper and Brignell, 1984). Considering further a differential capacitor sensor system, the two current generators are adjusted for identical times with the sensor(s) on null position. Both circuits will perform as defined by equation 9.16, so the system output or time difference in terms of the number N of clock pulses of frequency f, is given by,

$$\text{time} = N/f = \Delta C(V_{dd} - V_t)/I \tag{9.17}$$

where ΔC is the change in sensor capacitance.

A further process involves the use of the same principle as a dual slope converter. Consider Figure 9.17. With the switch in position 1 for a predetermined time (N clock pulses of period T), the integrator output V_i increases (as shown in Figure 9.17(b)). Thus,

$$V_i = \frac{NTV_r}{CR_1} \tag{9.18}$$

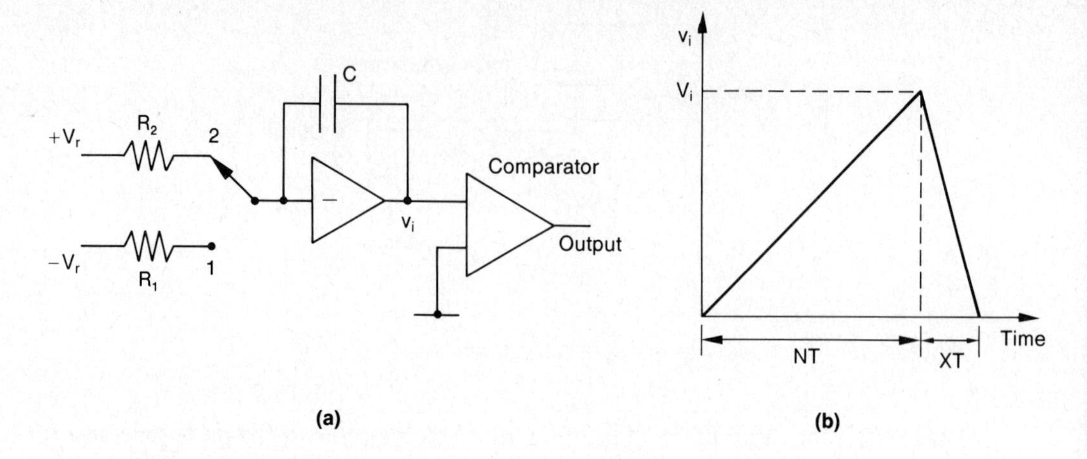

Figure 9.17 Principle of the dual slope analog to digital converter

After time NT, the switch is thrown to position 2 with the integrator output now decreasing linearly. When it reaches zero voltage, the comparator switches, indicating that the conversion process has finished. Thus,

$$V_i = \frac{XTV_r}{CR_2} \tag{9.19}$$

By equating with equation 9.18, the following is obtained,

$$X = \frac{NR_2}{R_1} \tag{9.20}$$

Resistors R_1 or R_2 can be made sensors so that equation 9.20 allows the change in sensor value to be determined. In Chapter 6 we saw that switched capacitors can be made to act like resistors, enabling capacitive as well as resistive sensors to be used.

9.4.3 Lateral inhibition

In many cases, a comparison of two or more signals is all that is required. Consequently, we have a differential system with two sensors feeding a simple comparator. Figure 9.18 gives two examples. In the first example, two diodes on the chip are used as temperature sensors. A large MOS transistor device mounted centrally is used as a heater. On reset, the bistable comparator is biased into an unstable state and held there by the charge on the diode junction capacity. Whichever diode is the hotter one, will have the larger leakage current to discharge the capacitor more quickly and so determine the final state of the comparator. Since the gas or fluid direction of flow determines which of the diodes is hotter, the comparator output is an indication of flow direction.

In a similar manner, the direction of the magnetic field determines the setting of the comparator (Figure 9.18(b)). Notice that if a known offset current is fed into

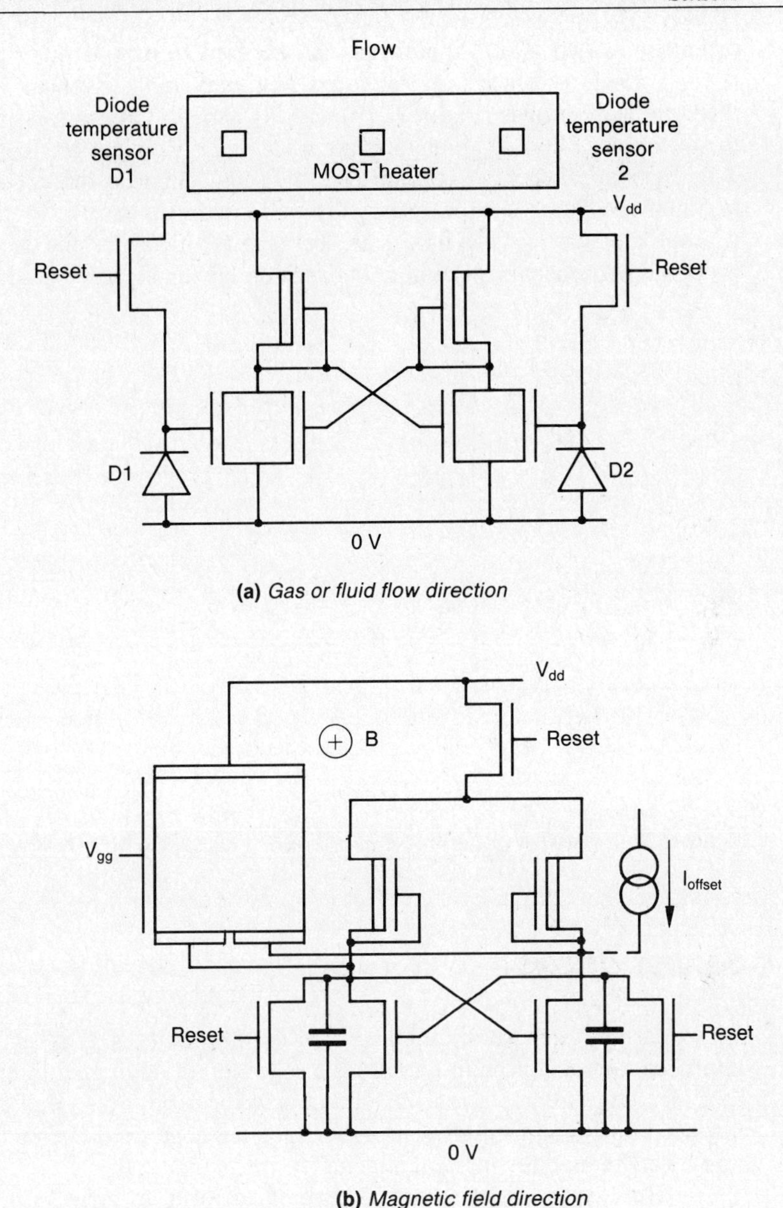

(a) *Gas or fluid flow direction*

(b) *Magnetic field direction*

Figure 9.18 Simple comparator sensors

one side, the comparator will only change state if the magnetic field is in the right direction and exceeds some threshold value. This is illustrated by the dotted line in Figure 9.18(b).

In many instances, a multistate system is required. Can the simple two state comparator be extended? In the late 1950s when transistors were expensive, decade counters were simply made by using 10-diode transistor NOR gates all cross connected in a ring so that at no time could more than one gate be ON. This same technique, called lateral inhibition, can be used to extend our simple comparator (Lyons, 1981). Consider an extension of the example given in Figure 9.18(a). If there are eight diodes (Figure 9.19) used as temperature sensors, we can determine the direction of flow to an angular accuracy of 22.5°. Eight NOR gates are required, with each one being cross connected. We can describe the process as taking an instantaneous look at the energy being sensed (on reset). Preprocessing occurs without any timing, the final state being determined by the energy sensed. The process is extremely powerful but depends on having an array of identical devices.

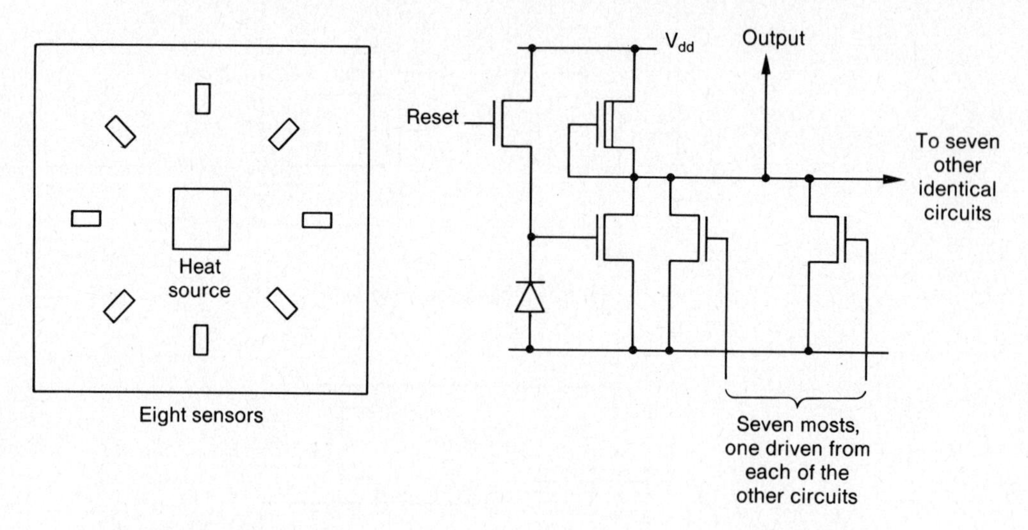

Figure 9.19 Lateral inhibition to determine the direction of glas flow

9.5 Sensor arrays

Sensor arrays are becoming increasingly important as they allow not only an improved rate of collection but also the collection of additional information. Sensors that are being used as arrays include optical for robot eyes, Hall effect for reading magnetic cards and pressure sensors for speech recognition. Even temperature sensor arrays can be used for thermal imaging.

How can sensors be interconnected to form an array? Consider the photo diode. Figure 9.20 shows a one dimensional array. The precharge charges the diode junction capacitor and any other built-in parallel. Like the comparator circuit discussed previously, the intensity of the light on the diode determines the rate of disccharge of the capacitor. At a preset instant in time, the charge left on the capacitor is to be sensed. This is achieved by using a shift register with a "1" circulating in it so that each capacitor is sampled in turn. The output is an analog voltage which can be processed by any of the methods discussed in Section 9.4.

One problem with this system is that the capacitors are all sampled at slightly different times. This can be overcome by including a second bank of intermediate storage capacitors so that all sensor capacitors are sampled simultaneously, after which the charge on the storage capacitors can be read out sequentially, using the

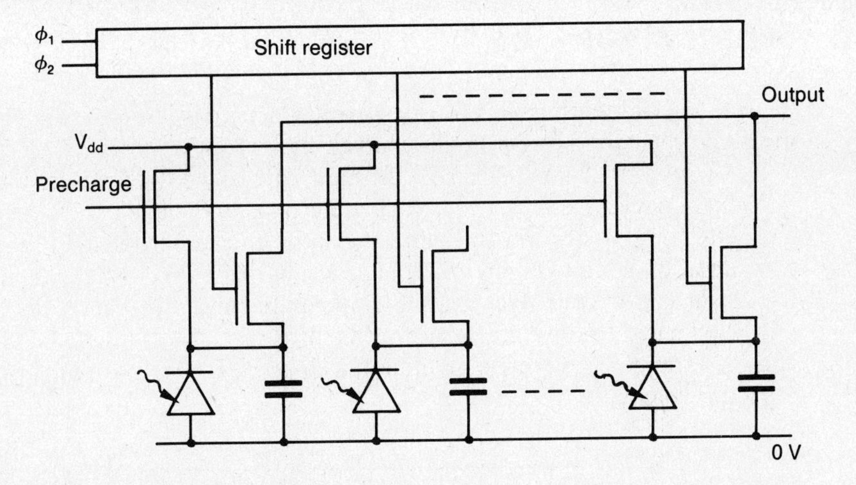

Figure 9.20 Simple 1D sensor array

shift register. Figure 9.21 illustrates this for a two dimensional array. Here, each sensor in a given row is read simultaneously, but each row is sampled at a different time.

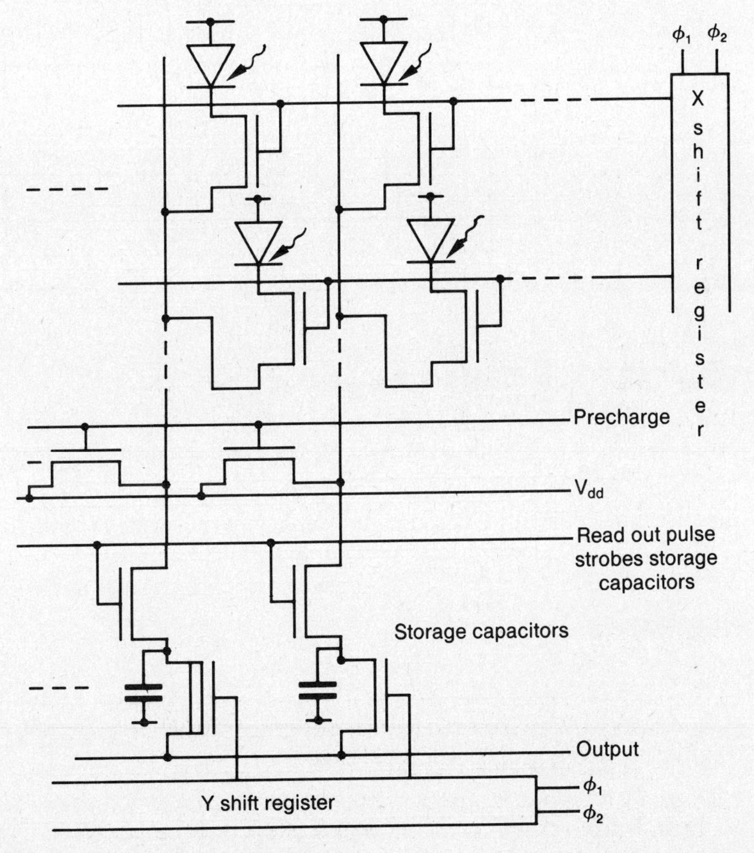

Figure 9.21 Two dimensional sensor array

9.6 Sensor chips

Several sensor chips have been designed which make use of the silicon sensor measurements, given earlier in this chapter.

Figures 9.22 and 9.23 show the basic circuit and floor plan of the design of a differential capacitive sensor, employing the discharge method. The design provides a digital output for feeding directly into a microprocessor. A temperature compensation circuit (Figure 9.22) has been added to the threshold circuit so that the sensor can operate over a wide temperature range. To cancel the effect of the unwanted metal to substrate capacitor, a polysilicon layer is inserted under the metal and fed through a unity gain amplifier (such as BUFAMP) from the on chip metal capacitor voltage.

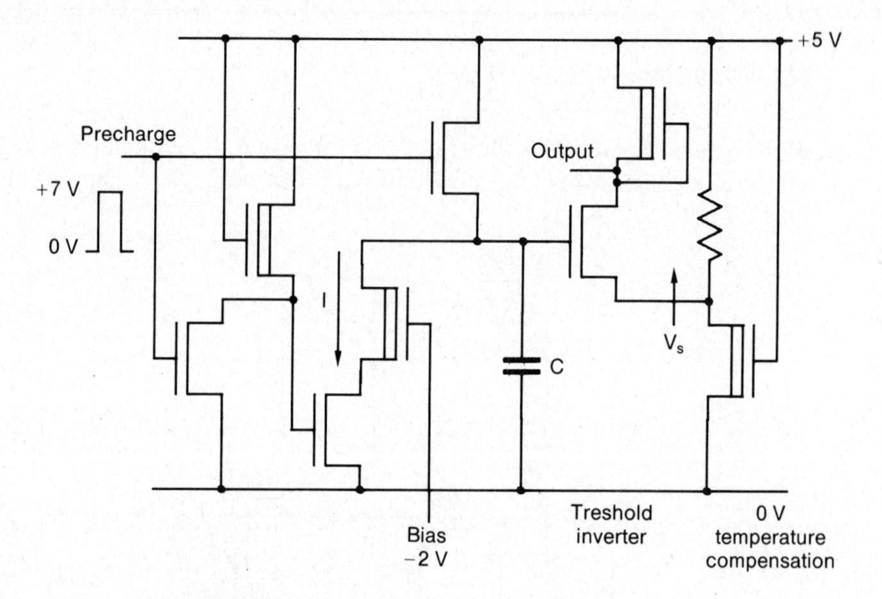

Figure 9.22 Temperature compensation of a threshold circuit

Figures 9.24 and 9.25 show a simple temperature and light monitor chip. Not being a differential system, the circuit is dependent upon the absolute value of diode leakage current. Consequently, each chip must be individually calibrated.

9.7 Exercises

1 Figure 9.19 shows eight diodes to detect the direction of heat flow, using lateral inhibition. Using stick diagrams, design an NOR logic system for the case of a 10-diode heat flow direction detector.

2 Figure 9.20 shows a one dimensional photo array where the diodes are accessed sequentially. Devise a system where all diodes are accessed at the same instant, but the data is still fed out sequentially.

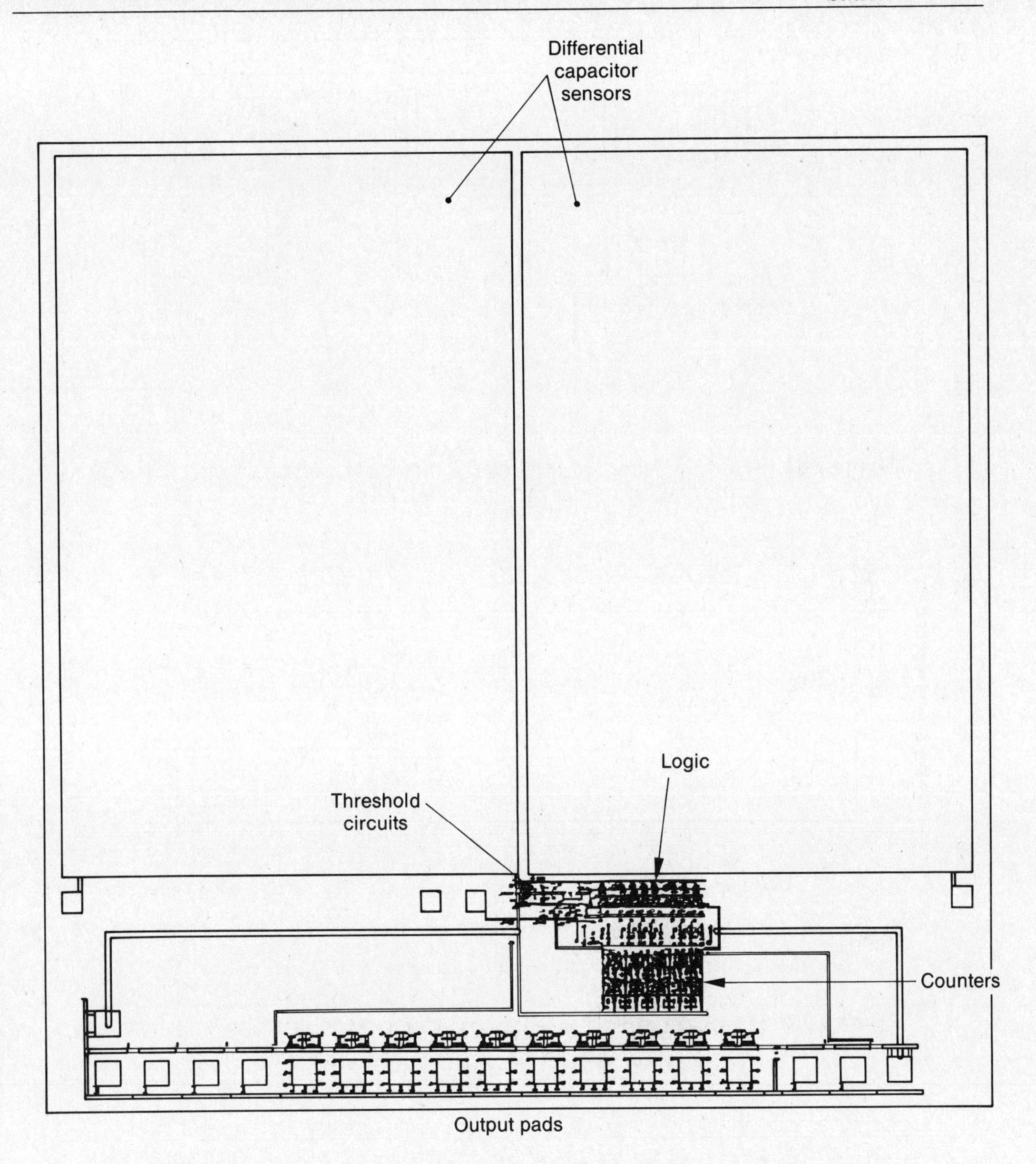

Figure 9.23 Layout of the differential capacitance ensor (metal layer only)

3 In the metal capacitor sensor shown in Figure 9.11, the parasitic capacity to substrate degrades performance. It is suggested that placing a polysilicon layer under the metal and driving it at the metal plate potential through a unity gain buffer amplifier will eliminate this problem. Show whether or not this is true.

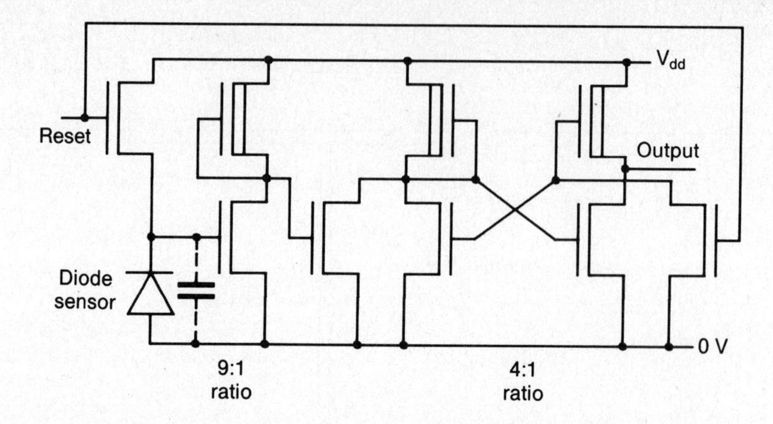

Figure 9.24 A simple current to time circuit that has been employed to measure both temperature and illumination level

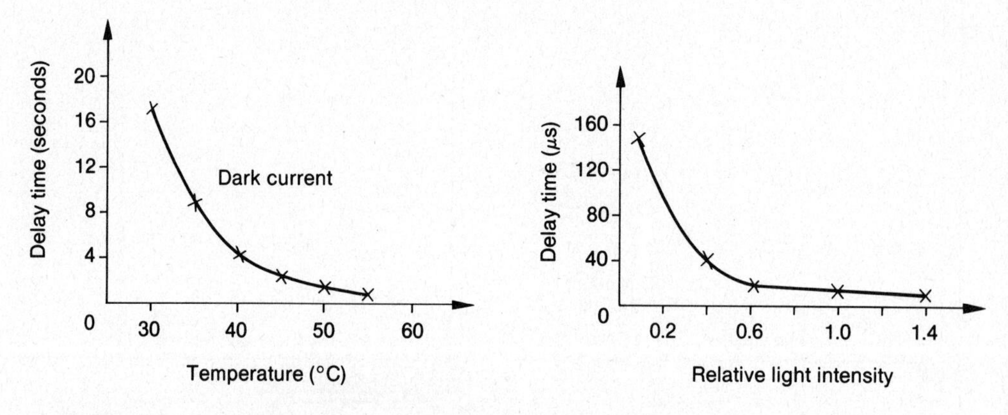

Note:
Curves for manufacturer 2 are similar in shape but are times three orders lower.

Figure 9.25 Temperature/light intensity monitor for Manufacturer 1

9.8 References

Blauchild, R. A., Tucci, P. A., Muller, R. S. and Meyer, R. G. (1978), "A New nMOS Temperature Stable Voltage Reference", *IEEE Journal of Solid State Circuits*, Vol. SC-13, pp. 767–74.

Braun, R. J., Chai, H. D. and Ebert, W. S. (1974), "FET Hall Transducer with Control Gates", *IBM Technical Disclosures Bulletin*, Vol. 17, No. 7, pp. 1895–6.

Chien, C. L. and Westgate, C. R. (eds), (1980), *The Hall Effect and Its Applications*, Plenumen Press, New York.

Cooper, A. R. and Brignell, J. E. (1984), "A Magnetic Field Transducer with Frequency Modulated Output", *Journal of Physics Education Science Instrum.*, Vol. 17, pp. 627–8.

Davies, L. W. and Barnicoat, G. P. (1970), "Integrated Hall Current Element", *Procedures of International Conference on Microelectronics Circuits and System Theory*, Sydney, pp. 32–3.

Green, M. A. (1982), *Solar Cells: Operation Principles, Technology and Systems Applications*, Prentice-Hall, Englewood Cliffs, NJ.

Hamilton, D. J. and Howard, W. G. (1975), Basic Integrated Circuit Engineering, McGraw-Hill, New York.

Haskard, M. R. (1983), *Sensors on nMOS Multi Project Chips*, Proc. IREECON 83, Sydney, 5–9 September 1983, pp. 15–17.

Haskard, M. R. (ed.) (1984), *Silicon Capacitive Sensor Programme: 1984*, Internal Report, Microelectronics Centre, South Australian Institute of Technology, June 1984.

Hoffman, C. R., Haskard, M. R. and Mulcahy, D. E. (1984), "Carbon Filled Polymer Paste Ion-selective Probes", *Analytical Letter*, Vol. 17, No. A13, pp. 1499–1509.

Kaiser, W. A. and Proebster, W. E. (eds) (1980), *From Electronics to Microelectronics*, North-Holland Publishing Co.. Amsterdam, pp. 648–50.

Lyons, R. F. (1981), *The Optical Mouse*, Xerox Corporation Report.

Meijer, G. C. M. (1982), *Integrated Circuits and Components for Band Gap References and Temperature Transducers*, Electronics Research Laboratory, Delft University of Technology, Delft, The Netherlands.

Regtien, P. P. L. (ed.) (1978), *Modern Electronic Measuring Systems*, Delft University Press, Delft, The Netherlands.

Window, A. L. and Holister, G. S. (1982), *Strain Gauge Technology*, Applied Science Publishers, Essex, UK.

10 Ancillary circuits

10.1 Types of ancillary circuits needed

Where analog signals are to be processed on a chip, analog pads are required to allow signals to flow on and off the chip. There are, however, other occasions where nondigital pads are required for probing, monitoring, setting reference signals, and so on. Consequently, a set of analog pads, shown diagrammatically in Figure 10.1, has been designed. Appendix B includes more detailed layouts.

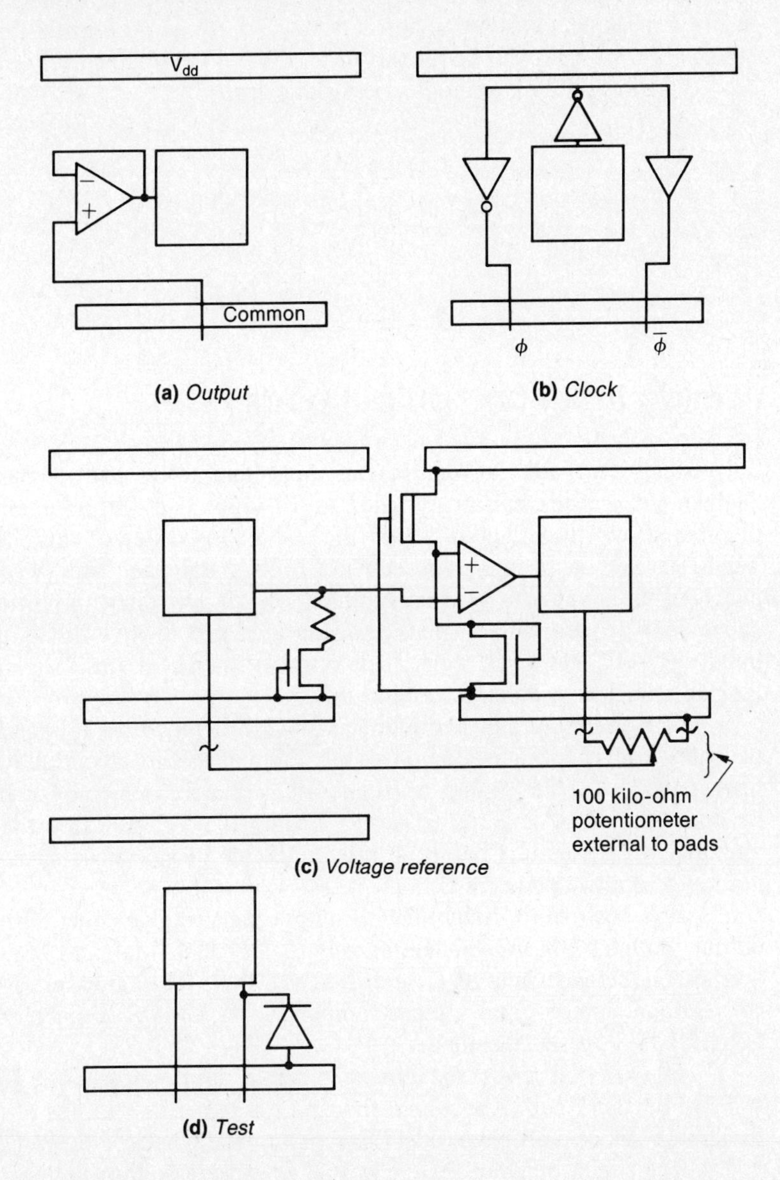

Note:
The input pad is the digital input pad.

Figure 10.1 Pads

Because switched capacitor methods are recommended for MOS analog work, digital pads are also required for the digital clock lines, with super buffers being needed to drive long clock lines. The tiles for such amplifiers are included in Appendix B.

Finally, because the chip fabrication time can take several months, it is often essential to organize an experimental design so that the maximum amount of information can be derived from a chip. This is particularly so with an undergraduate course where there is no time for a second fabrication run to correct errors. Consequently, some time will be devoted to the problem of how probing and connection methods can be employed to allow on chip surgery, for the purpose of recovering information on poorly designed chips.

10.2 Analog input and output pads

Surprisingly, the conventional digital input pad (Hon and Sequin, 1980) with "lightening" arrester can be retained as an input pad. While it inherently has a distributed low pass filter in series, the cut off frequency of this filter is typically 125 MHz and is of little consequence. There may be cases when neither the increased shunt capacity nor the extremely large input voltage protection are wanted and a simpler pad, with a small diffusion area tied to it to discharge any voltage build up, is all that is necessary. It is possible to employ this simple pad for other functions, such as probing or testing.

The analog output pad is much more difficult. First, it is desirable that the pad runs from the standard 5 volt supply rail and that its layout allows both supply lines (0 to V_{dd}) to be routed, as is the usual practice, around the chip periphery. Second, since it is possible to design analog filter chips using the conventional amplifiers discussed in Chapter 4, it is preferable to have an output pad that does not required a two phase clock.

It is possible to design nMOS output pads to give either current or voltage output. Both have been included, namely PADOUT I (May and Macintosh, 1982), a transconductance amplifier, and PADOUT V (Haskard and May, 1984), a conventional unity gain voltage amplifier. A CMOS version of the latter, CPADOUT V, is also included.

Figure 10.2 gives the circuit diagram and performance figures of the transconductance pad. Notice that the circuit is dependent on the absolute value of R. Should this be a problem, then this resistor must be placed externally.

The circuit diagram for the nMOS unity gain voltage buffer pad PADOUT V is given in Figure 10.3, with typical performance figures listed in Table 10.1. Figure 10.4 and Table 10.2 give similar information for the CMOS analog output pad.

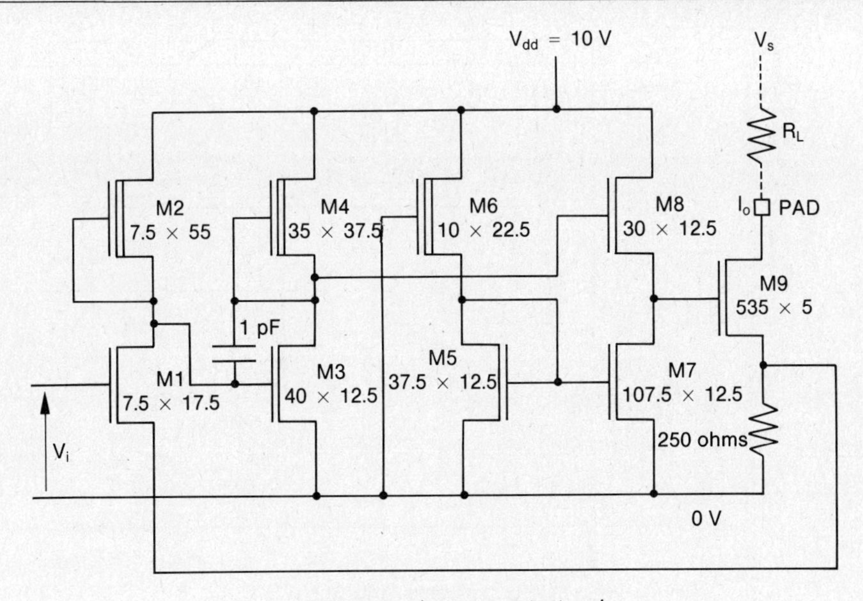

(a) *Transconductance output pad*

Note:
Dimensions are given in microns with width given first.

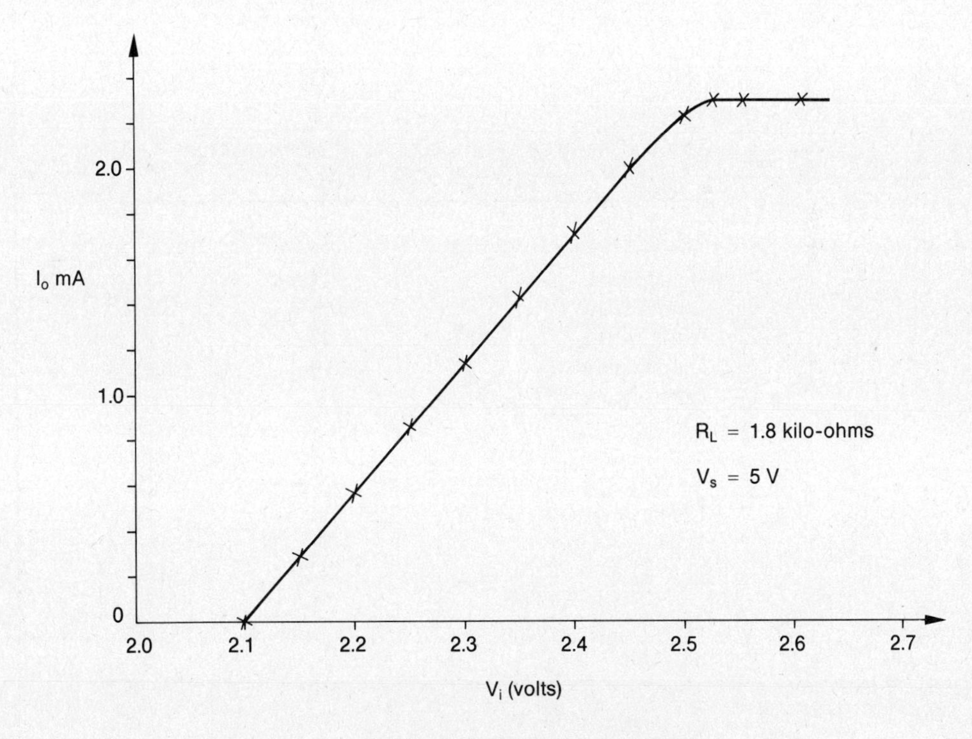

(b) *Typical transfer characteristics*

Figure 10.2 PADOUT I

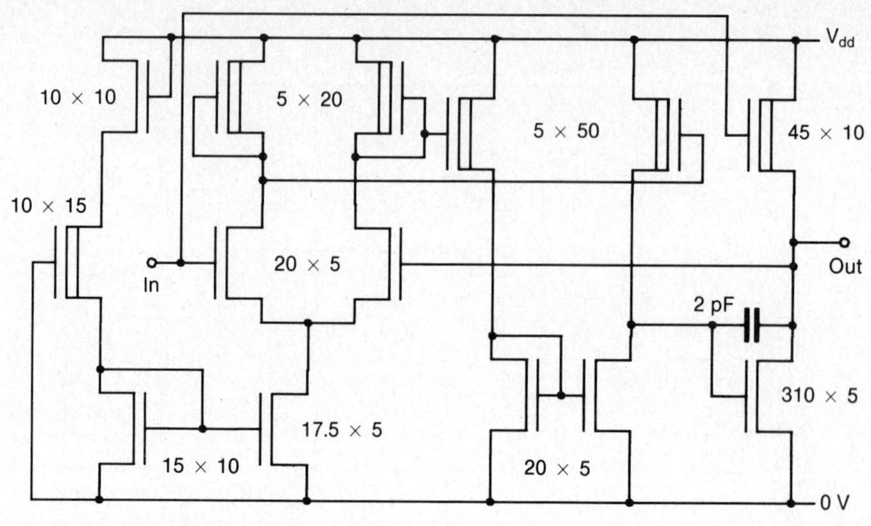

Notes:
1. Dimensions are in microns, with width included first.
2. V_{dd} = 5 volts.

Figure 10.3 Circuit diagram PADOUT V

Table 10.1 Measured closed loop performance parameters of PADOUT V

Parameter	*Measured value*	*Units*
Voltage gain	.988	—
Output swing	0.6–4.1	volts
Offset voltage (V_i = 2.5V)	17.6	mV
Output impedance	110	ohms
Cutoff frequency (−3 db)		
C_{LOAD} = 20 pF	5.7	MHz
C_{LOAD} = 100 pF	4.1	MHz
C_{LOAD} = 1000 pF	1.6	MHz
Power bandwidth (−3 db)		
(0 db = 3V p–p)		
C_{LOAD} = 20 pF	2.0	MHz
C_{LOAD} = 100 pF	1.6	MHz
C_{LOAD} = 1000 pF	.27	MHz
Slew rate		
C_{LOAD} = 20 pF	10	V/μS
C_{LOAD} = 100 pF	8.3	V/μS
C_{LOAD} = 1000 pF	1.7	V/μS

Note:
Supply voltage = 5V.

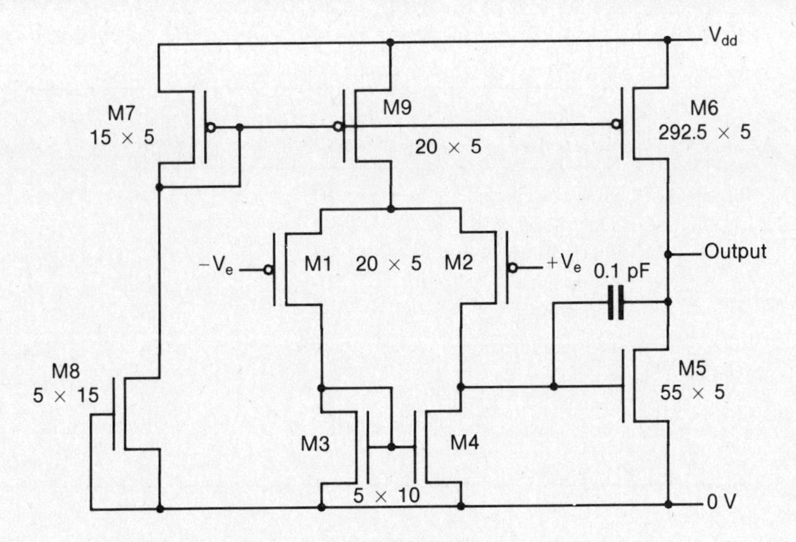

Figure 10.4 CMOS analog output pad CPADOUT V

Table 10.2 Performance figures for CPADOUT V

Parameter	*Measured value*	*Units*
Voltage gain		
Open loop	1300	—
Closed loop	0.999	—
Output swing	0.5–4.2	volts
Output impedance	50	ohm
Offset voltage	10	mV
Cutoff frequency (−3 db)	6	MHz
Supply current	300	μA

Note:
V_{dd} = 5V.

10.3 Other pads for analog work

For analog to digital conversion work, it is necessary to have a known stable voltage reference. Included within PADOUT V is a constant voltage/current source, with the circuit being adaptable in order to provide a simple voltage reference. Table 10.3 gives the spread in values for this reference voltage, showing how it varies with process parameters. Should the magnitude of the voltage be noncritical, PADOUT V can be used as it stands, with the voltage being taken from the gate of the differential pair current source.

Where the voltage magnitude is important, two pads are brought out so that, with an external 100 k ohm potentiometer, the reference voltage value can be

Table 10.3 Performance figures for the nMOS constant voltage reference pad PADREF

Pad type	*Parameter*	*Measured value*	*Units*	*Comments*
Output	Standby current	450	μA	Design value
	Open loop gain	12–53	—	
	Output impedance	—	—	Design value 40–80
	Open loop	10	kilo-ohms	
	Closed loop	300	ohm	
Voltage reference	Internal reference	1.65–1.92	volts	Range on 10 samples
	Range	2.35–2.54	volts	Sample 1, set 2.5V, using fixed resistors Range, using same resistors for 10 samples

accurately set. Figure 10.5 gives the circuit diagram of variable voltage reference PADREF, while Appendix B contains the layout.

As explained previously, switched capacitor techniques require not only a two phase nonoverlapping clock, but often the complement of the clock to achieve charge neutralization on the input of high gain amplifiers and comparators. PADCLOCK is a pad that takes a clock input, buffers it and simultaneously provides the clock complement. It is included in Appendix B.

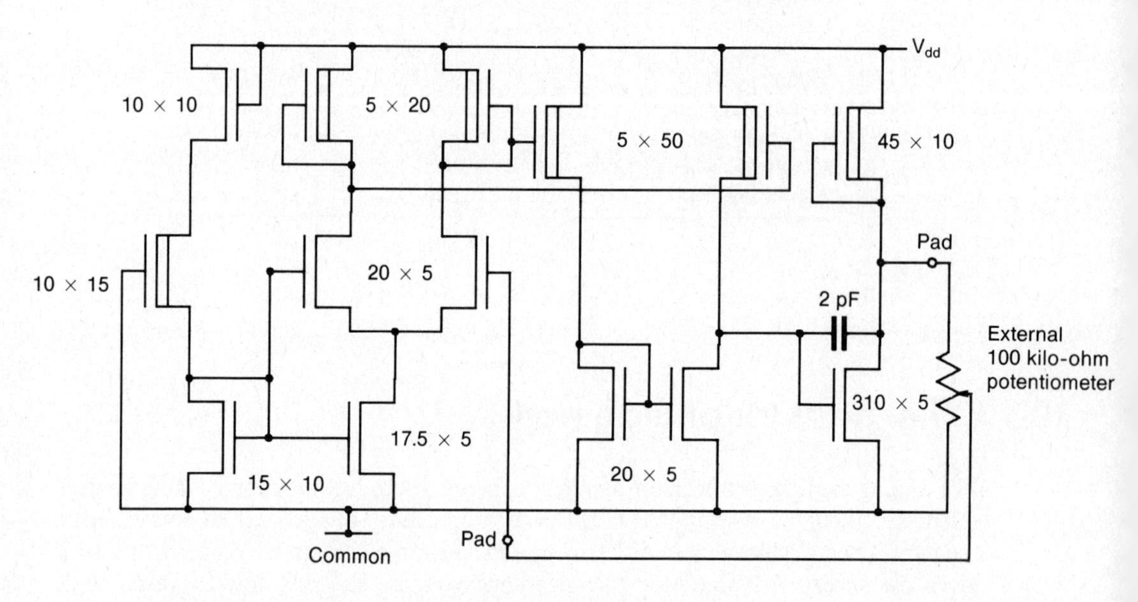

Notes:
1. Dimensions are in microns with width given first.
2. V_{dd} = 5V.

Figure 10.5 PADREF: A "variable" constant voltage pad

10.4 Pads to assist in testing chips

Examination of multiproject chip designs suggests that approximately 50 percent will have errors of some kind. Table 10.4 summarizes the errors from Australia's multiproject chip run AUSMPC 5/82.

Since all chips were design rule checked, all of these faults have passed through a geometric checker. It therefore seems practical to incorporate in any layout means of testing, isolating and even possibly repairing faulty sections. To do so, the layout design rules need to be modified and a range of microelectronic repair equipment procured. With regard to the latter, this would include a probing station with microscope, low capacitance probes and ultrasonic cutting probe (both with suitable manipulators) and silver loaded conducting epoxy. An ultrasonic or thermosonic bonder is also very useful.

Table 10.4 Summary of errors from multiproject chip AUSMPC 5/82

	Logic/control error				*Clock error*		*Supply line error*	
Chip level of success	Design error	Incorrect layout connection	Omitted layout connection	Layout short to supply line	Design error	Incorrect layout connection	Omitted common or V_{DD}	Short common to V_{DD}
Working 80 percent or better	4	4	2	—	2	1	—	—
Major fault	5	1	1	3	1	—	2	2

The following two modifications of design rules, are suggested. First, the modules or mosaics are not to be abutted, but there is to be a space left between them to allow isolation of that circuit. These spaces or separations are to have no overglaze, for in these regions additional probing pads to observe and control, exposed tracks for microcutting, fusable links for blowing and lastly joining pads are to be included. This concept is illustrated in Figure 10.6.

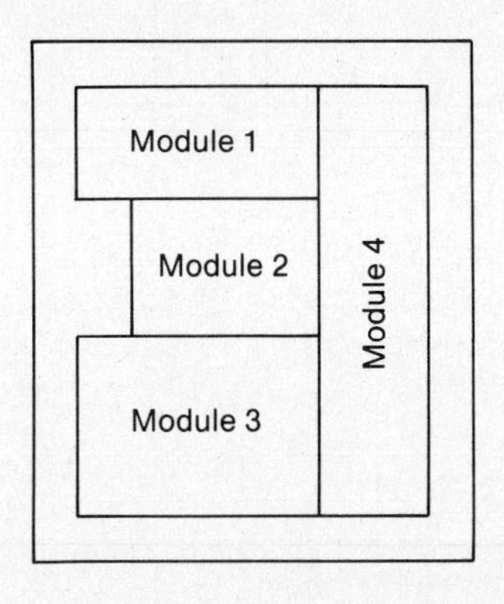

(a) *Normal abutting cells*

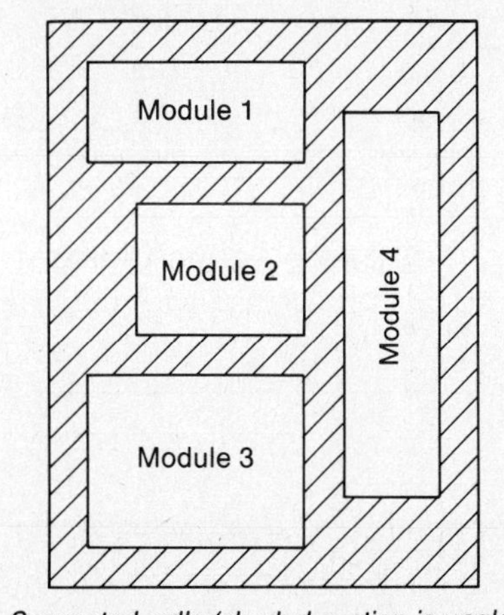

(b) *Separated cells (shaded section is unglazed)*

Figure 10.6 Concept of separated cells with isolation/joining/testing structures in between

The separation allows the insertion of small pads, which leads on to the second design rule change, that is, the inclusion of a pad library for observing, controlling, separating and joining. Figure 10.7 summarizes a proposed pad library at the Microelectronics Centre, at the South Australian Institute of Technology. An additional concept is that of removing the glaze from the top of capacitors for it can often aid testing, as voltages on the metal plate of the capacitor can be probed.

While probe tips that allow probing onto 7.5 μm metal tracks exist, it is very easy to damage and open circuit the aluminium. Further since it is necessary to bring to the surface polysilicon and diffusion layers which need to be probed, it is suggested that all these pads be of modified contact hole dimension, giving a 30 μm square metal pad for probing. To observe waveforms at these points, the probes used must be of low capacity (high impedance) containing super buffers. The concept of observing is not at all difficult, however these pads can be used for a second function, namely controllability. This is a much more difficult concept, for here one is endeavoring to set a particular point in a circuit to a desired voltage or logic level. With inverter type circuits, only simple rules need to be followed, for the circuit is often selfprotecting. Consider the case of a simple nMOS inverter (or NOR gate), as shown in Figure 10.8. If the pad is on the output, by shorting the pad to ground with a resistive short (e.g. a simple short circuit) for a low output, or to V_{DD} for a high output (again a resistive short), the inverter will not be damaged and the input into the circuit the pad is driving will be controlled. With analog circuits, similar techniques can be employed, except the resistive shorts will normally be of non-zero resistance.

To isolate cells, tracks need to be severed. This is simplified if the probing station contains an ultrasonic microcutter. If it does not, then fusable links in metal can be used. Here the tracks to be cut must be brought to the metal layer and, using the pads suggested in Figure 10.7, the small minimum width track can be blown

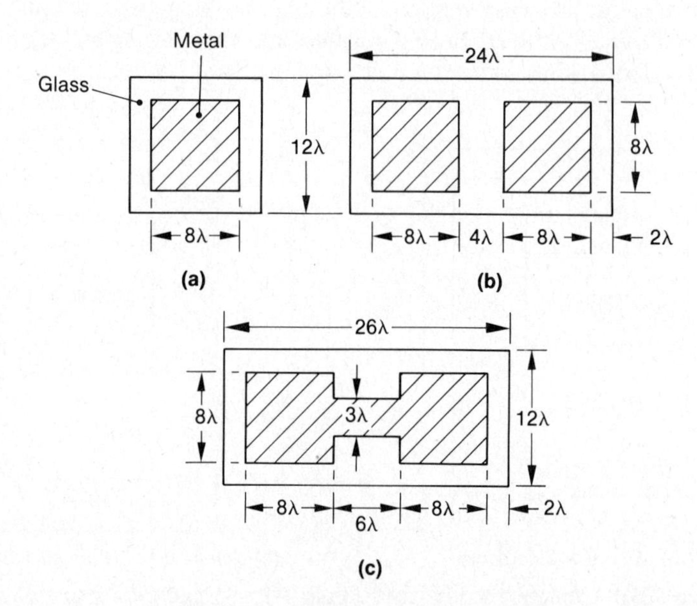

Figure 10.7 Summary probe, joining and separating pads

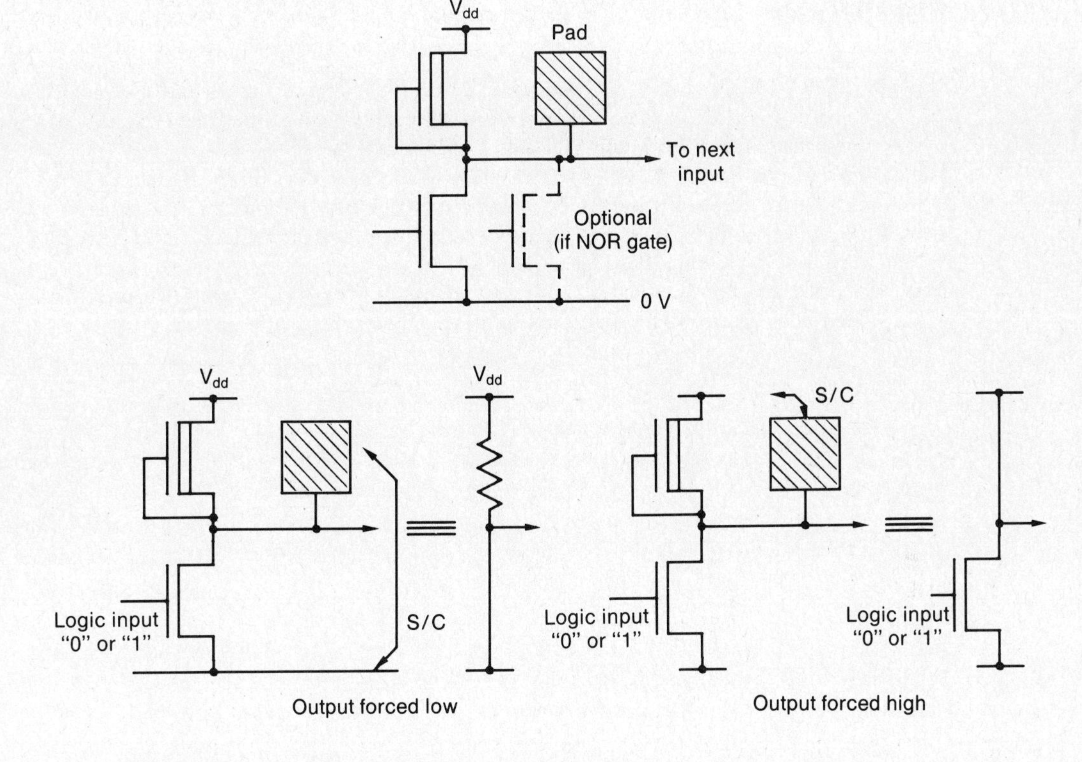

Figure 10.8 Example of a pad to control an input

with a 0.5 amp current discharge from a capacitor or DC power supply. A short fusable link is used as it provides a more repeatable fusing than does a long fuse.

The final operation is the planned joining of tracks. Here it is suggested that two metal pads that need joining are spaced at minimum spacing to allow silver epoxy glue to be smeared across the gap. As with all repairs, the work must be carried out carefully under a microscope.

10.5 Exercises

1 A transconductance output pad is required to accurately provide a transconductance of 5 mA/V. Devise a layout for the circuit given in Figure 10.2, where R is external to the chip.

2 If several of the above transconductance amplifiers are required on a chip, is it possible to modify the circuit so that only a common single external resistance sets the transconductance value? If so, how would you do this?

3 A noncritical internal constant reference voltage is required. It must be between $1\frac{1}{2}$ to 2 volts. How would you use PADOUT V (Figure 10.3) in this application?

10.6 References

Haskard, M. R., Marriage, A. J. and Bannigan, J. T. (1983), "Salvaging Silicon—Proposals for Repairing MPC Projects", *Conference Digest*, Creating Integrated Systems—An Australian Silicon Workshop, Adelaide, 23–25 May 1983, pp. 33–8.

Haskard, M. R. and May, I. C. (1984), "A Library of Analogue Cells for the System Designer", *Conference Digest*, VLSI–PARC Conference, Melbourne, 15–17 May 1984, pp. 65–6.

Hon, R. W. and Sequin, C. H. (1980), *A Guide to LSI Implementation* (2nd ed.), Xerox, Palo Alto Research Centre, Palo Alto, Cal., SSL–79.

May, I. C. and Macintosh, C. (1982), *Transconductance Amplifier*, Internal Report, Microelectronics Centre, South Australian Institute of Technology, Adelaide, SA.

11 Transistor circuit modeling

11.1 Models of MOS transistors

The simple model developed in Chapter 2 could suggest that the MOS transistor is well understood and easy to model. However, new models are continually being developed and, as the dimensions of devices shrink, an ever-increasing number of new effects becomes significant which must be included in the model. (For further reading on new models, see Hussan et al., 1982; Kumar, 1980; McCreary, 1983; Menckel et al., 1972; Liu and Nagel, 1982; Shrivastava and Fitzpatrick, 1982; Silburt et al., 1984; Simard-Normandin, 1983; and Vladimirescu, 1980.)

Why is a model required at all? Gone are the days when one could go to a store, collect a handful of components and wire them up to test an idea fully. With integrated circuits, there are additional parasitic components, and stray reactances of breadboard systems can be too large in value. Consequently, the breadboard system does not accurately represent the final circuit. Other factors such as component matching and total numbers of components also make breadboarding impossible. Computer simulation methods must be used in order to validate designs. Hence, an accurate mathematical model of a transistor is required. The more accurate the model the better the computer prediction. The simulation of circuits using computers is in keeping with present day computer aided engineering techniques and expert systems. The total design can now be accomplished (including simulation, layout and verification) at a designer's computer work station. Naturally, the final test of these simulations is whether or not the chip performs as predicted.

For analog simulation, the SPICE program is universally used (Nagel, 1975). Over the years it has been upgraded and present versions allow the choice of three inbuilt MOS models (Vladimirescu, 1980; Alexander, 1977; Schichman and

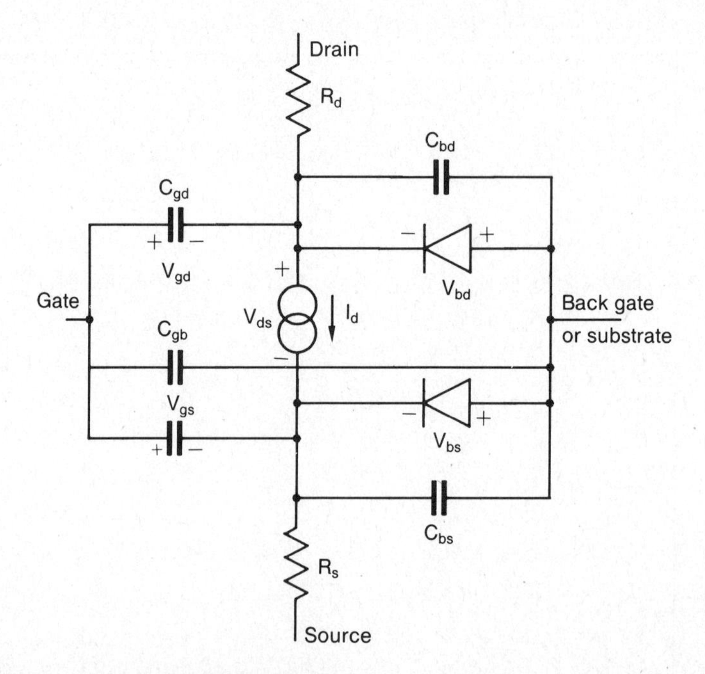

Figure 11.1 SPICE model of an nMOS transistor

Hodges, 1968). Fortunately with MOS multiproject chip work, process variations are so great that the predictions of even the simplest model in SPICE are adequate. Figure 11.1 shows such a SPICE model (Alexander, 1977).

Determining the parameters is difficult with any model, with many papers having been written on how to determine MOS model parameters (Anderson and Smith, 1979; Ward and Doganis, 1982; Haskard, 1983; Thoma and Westgate, 1984; Peng and Afromowitz, 1982). If it is too complex a process, the model tends not to be used. Fortunately, because of process variations, only six crucial parameters are needed for modeling MOS multiproject chip transistors, these being determined by a number of very simple measurements.

Before examining this method, we need to deal with the use of SPICE to model circuits. Experience indicates that the simplified SPICE model, level 2, provides excellent static or DC simulation of MOS circuits. With transient and AC simulation, program convergence may be difficult or time consuming, in which case it is best to calculate all circuit capacities and add them in as discrete capacities, independent of the MOS model. In this way, capacities in parallel can be combined, thereby reducing the complexity. This not only speeds up convergence, but often makes convergence possible when previously it could not be achieved.

A further trick that often solves DC convergence problems with older versions of SPICE is to add to the circuit dummy reverse biased diodes. They are positioned so as not to interfere with circuit operation.

11.2 Determining MOS transistor and model static parameters

Some 29 parameters are required to fully describe the SPICE model given in Figure 11.1 (Alexander, 1977). Fortunately, in the case of multiproject chip work only six are important for DC analysis, namely:

1. V_{t0}, the zero bias threshold voltage;
2. K (or KP), the intrinsic transconductance;
3. gamma (γ), the bulk threshold parameter;
4. phi (ϕ), the strong inversion surface potential;
5. lambda (λ), the channel length modulation parameter; and
6. LD, the lateral diffusion distance.

The channel modulation parameter lambda (λ), should not be confused with the scaling term used in previous chapters.

Consider the SPICE model for a transistor in the nonsaturation region with no substrate bias ($V_{bs} = 0$). Thus,

$$I_d = K\frac{W}{L}V_{ds}\left(V_{gs} - 2\phi_f - V_{fb} - \frac{V_{ds}}{2}\right) - \tfrac{2}{3}\gamma((V_{ds} + 2\phi_f)^{\frac{3}{2}} - 2(\phi_f)^{\frac{3}{2}}) \quad \textbf{(11.1)}$$

and,

$$V_{to} = V_{fb} + 2\phi_f + \gamma\sqrt{2\phi_f} \quad \textbf{(11.2)}$$

Combining and rearranging these equations, we obtain,

$$I_d = K\frac{W}{L}V_{ds}\left(V_{gs} - V_{to} - \frac{V_{ds}}{2}\right) - \tfrac{2}{3}\gamma((V_{ds} + 2\phi_f)^{\frac{3}{2}} - (2\phi_f)^{\frac{3}{2}} - \tfrac{3}{2}(2\phi_f)^{\frac{1}{2}}V_{ds}) \qquad \textbf{(11.3)}$$

The second term can be considered a constant, K_{ds}, if V_{ds}, the drain source voltage, is held constant, for both γ and ϕ_f are process constants. Thus when V_{ds} is constant,

$$I_d = K\frac{W}{L}V_{ds}\left(V_{gs} - V_{to} - \frac{V_{ds}}{2}\right) - \tfrac{2}{3}\gamma K_{ds} \qquad \textbf{(11.4)}$$

and rearranging these equations, we obtain,

$$V_{gs} = \frac{L}{KW}\frac{I_d}{V_{ds}} + \left(V_{to} + \frac{V_{ds}}{2} + \frac{2}{3}\frac{\gamma K_{ds}L}{V_{ds}KW}\right) \qquad \textbf{(11.5)}$$

This equation is in the form of the equation of a straight line (variables V_{gs} and I_d/V_{ds}), where the slope is L/KW and intercept $V_{to} + \frac{1}{2}V_{ds} + \frac{2}{3}\gamma K_{ds}L/(V_{ds}KW)$. If ϕ_f and γ are known, then for a transistor of given geometry (L and W), V_{to} and K can be determined. Because MOS fabrication processes have reasonably large parameter spreads, a constant value is assumed for $\phi_f = 0.325$. In practice, this has been found to be of sufficient accuracy so that only γ need be determined.

For an ideal transistor in the saturation region, a graph of the square root of drain current versus gate source voltage is a straight line. With real devices of small geometry, the graph is a slowly varying S shape. If a reverse substrate bias is applied, there is a lateral shift to the right of this curve. By noting this shift (ΔV_t), a value of γ can be calculated by solving equation 11.2, for threshold voltage with no substrate bias, and the following equation, for the SPICE model with an applied substrate bias:

$$V_t = V_{fb} + 2\phi_f + \gamma\sqrt{2\phi_f - V_{bs}} = V_{to} + \Delta V_t \qquad \textbf{(11.6)}$$

The final parameter λ can be determined as follows:
In the saturation region, the SPICE model can be approximated by,

$$I_d = \frac{K \times W(V_{gs} - V_{to})^2}{L(1 - \lambda V_{ds})} \qquad \textbf{(11.7)}$$

If V_{gs} is held constant then the drain current for two different drain source voltages are related by,

$$\frac{I_{d1}}{I_{d2}} = \frac{1 - \lambda V_{ds2}}{1 - \lambda V_{ds1}} \qquad \textbf{(11.8)}$$

from which λ can be calculated.

These SPICE model parameters can be determined by the following three simple measurements in Figure 11.2 for an n-channel enhancement mode transistor.

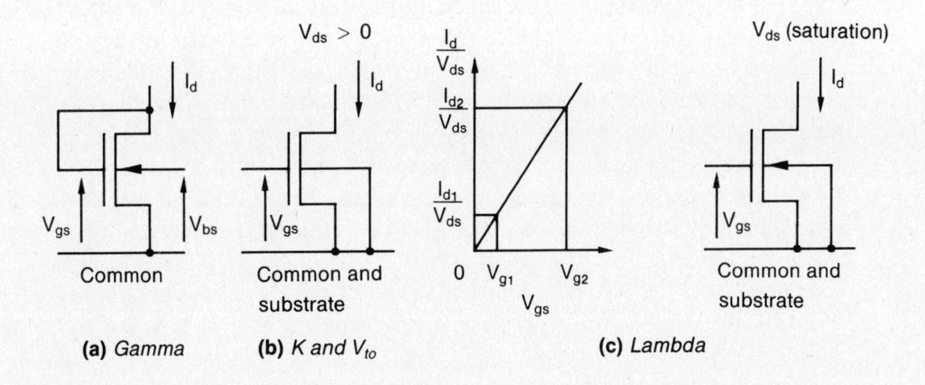

Figure 11.2 Simple test circuits for determining gamma, K and V_{to} and lambda

1. Determination of gamma (γ). Refer to Figure 11.2(a). For $V_{bs} = 0$, $V_{ds} = V_{gs} = V_1$ measure the drain current I_d. Set $V_{bs} = -V_2$ and restore the drain current to the original value by adjusting V_{gs} to V_3. Thus,

$$\gamma\sqrt{V_2 + 0.65} = \gamma\sqrt{0.65} + V_3 - V_1 \qquad \textbf{(11.9)}$$

Solve for γ.

2. Determination of K and V_{to}. Refer to Figure 11.2(b). For preferably a large transistor, so that channel shortening effects are insignificant, set V_{ds} to a near zero voltage to ensure that the device is in the nonsaturated region. For two values of gate source voltage (V_{g1} and V_2), note the drain current (I_{d1} and I_{d2}).

Thus,

$$K = \frac{L}{W} \times \frac{(I_{d2} - I_{d1})}{V_{ds}(V_{g2} - V_{g1})} \qquad \textbf{(11.10)}$$

and,

$$V_{to} = V_{g1} - \left\{\frac{I_{d1}L}{V_{ds}WK}\right\} - \left\{\frac{V_{ds}}{2}\right\} - \left\{\frac{2}{3}\gamma\frac{K_{ds}L}{V_{ds}KW}\right\} \qquad \textbf{(11.11)}$$

where,

$$K_{ds} = (V_{ds} + 0.65)^{\frac{3}{2}} - (0.65)^{\frac{3}{2}} - (0.65)^{\frac{3}{2}} - \tfrac{2}{3}(0.65)^{\frac{1}{2}}V_{ds} \qquad \textbf{(11.12)}$$

3. Determination of lambda (λ). Refer to Figure 11.2(c). Since the value of lambda varies with device size, it is preferable to use a device of a similar size as those to be used in the final circuit. For a gate source voltage V_4 and two values of drain source voltage (V_5 and V_6), the first being just after the onset of saturation: note the two corresponding values of drain current (I_5 and I_6). Using equation 11.8, solve for lambda.

When performing these calculations, phi has been assumed to be equal to 0.65. Should the SPICE model using these computed parameters not provide a satisfactory match with measured static characteristics, the value of phi can be changed and calculations repeated until the desired match is achieved. In practice, this has not been found to be necessary.

Finally, manuals on SPICE give the lateral diffusion parameter LD as a ratio. For some of the programs examined, however, it is an absolute value. Measurement 2, made on devices of two different geometries, can be used to determine a value for LD, assuming that there are no photolithographic errors. Alternatively, if the junction depth of the source drain diffusions is known (typically 0.6 μm for γ = 2.5 μm processes), a practical value for LD is 0.8 of the junction depth.

Using the above method, SPICE parameters for both enhancement and depletion transistors can be determined.

11.3 Determining bipolar transistor model parameters

The bipolar transistor is a far more complex device to model than the MOS transistor. Consequently, models such as the Gummel–Poon, as used in the simulation program SPICE (Figure 3.8), require some 40 parameters. Many of these parameters are difficult to determine. Getreu, 1976, however, provides a detailed description of how the parameters may be determined. In some cases, several methods are given so that, depending on whether the device is to be employed in a pulse or analog circuit, the most appropriate measurement technique can be selected. This is satisfactory for determining standard discrete transistor parameters, but is not satisfactory for integrated circuit design where the designer has complete flexibility as to the geometry of the transistors. Fortunately, because device parameters have large spreads, typically -50 - $+100$ percent, it is not necessary to consider all the parameters or to become too involved in determining the appropriate measurement technique.

Six important parameters need to be determined. Because the bipolar transistor, unlike its MOS counterpart, is not symmetrical in construction, two sets of parameters are required—one in the normal (forward) mode of operation and the other in the inverted (reverse) mode. In addition to these, bulk resistance parameters (rb′, rc′ and re′) may be important, but these can be best estimated by the designer, using standard sheet resistance calculations. Thus the important parameters that need to be determined by measurement for minimum emitter geometry devices are:

1. the saturation leakage current;
2. the common emitter current gain;
3. the Early voltage;
4. the high knee current where high current injection effects become dominant;
5. the two parameters to describe the change in characteristics at low current, due to recombination and other effects.

Each of these parameters must be determined for the transistor connected in its inverted as well as normal mode of operation.

The common emitter current gain is determined by setting up the transistor as shown in Figure 11.3. Using incremental changes in base current, the change in collector current is noted and current ratio is calculated.

Measurements are made at the desired collector emitter voltage. The ambient temperature should also be recorded as current gain increases with temperature.

The Early voltage effect relates to how well the transistor approximates a constant current generator. The method of measurement is shown in Figure 11.4. By changing the collector emitter voltage and noting the collector current change for a fixed base current, the slope and, then, projected voltage of the intersection of the characteristic and the horizontal axis can be calculated. This is the Early voltage. Unfortunately, as shown in Figure 11.4, not all characteristics intersect at the same point and several measurements need to be taken to obtain an average value for this voltage.

The remaining parameters can be determined from one measurement. A plot is made of the collector current versus base emitter voltage for the base collector voltage zero. This is perhaps the most difficult of all the measurements because to obtain accurate values for the saturation current and two low current parameters

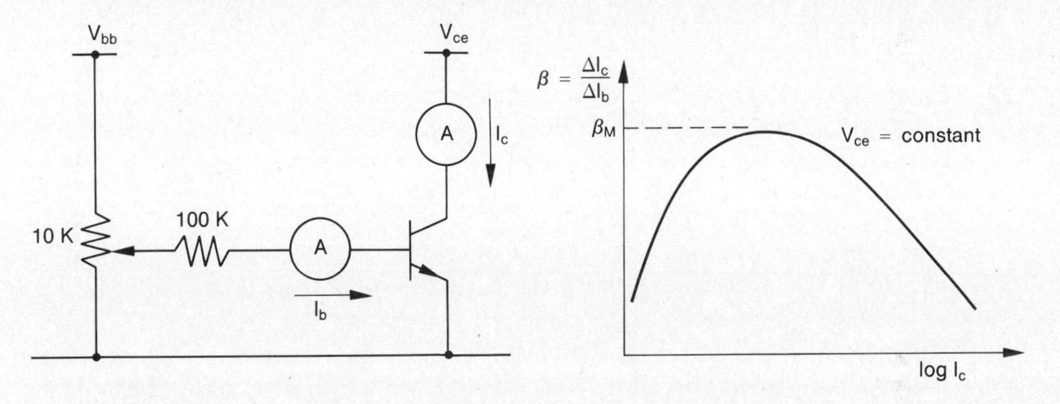

Figure 11.3 Determination of forward (or reverse) current gains

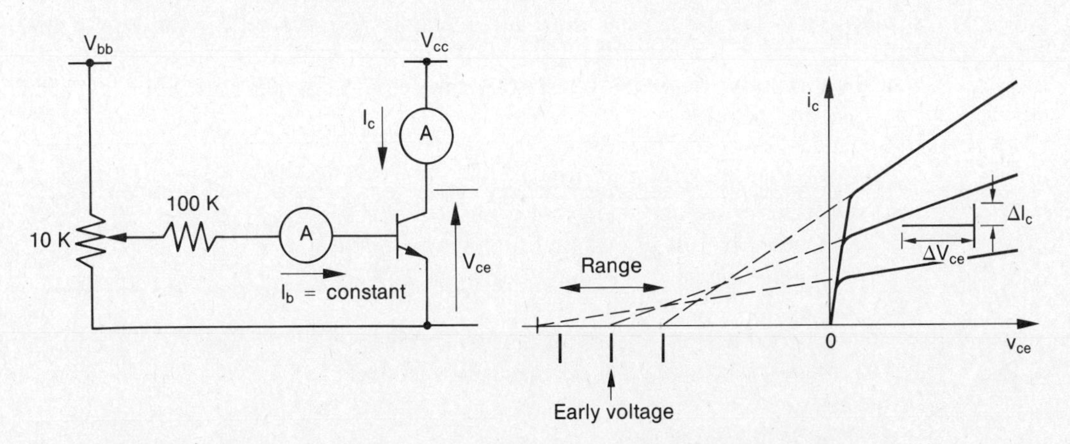

Figure 11.4 Determination of the forward (or reverse) Early voltage

(C_2, N_E forward; C_4, N_C reverse), measurements must be made at current levels, down to a picoamp. Depending on how noisy the environment is, these measurements may have to be made in a shielded box. Figure 11.5 shows the measurement circuit and how the parameters can be extracted.

The remaining parameters of capacity and resistance can all be calculated in the normal way, using device geometry.

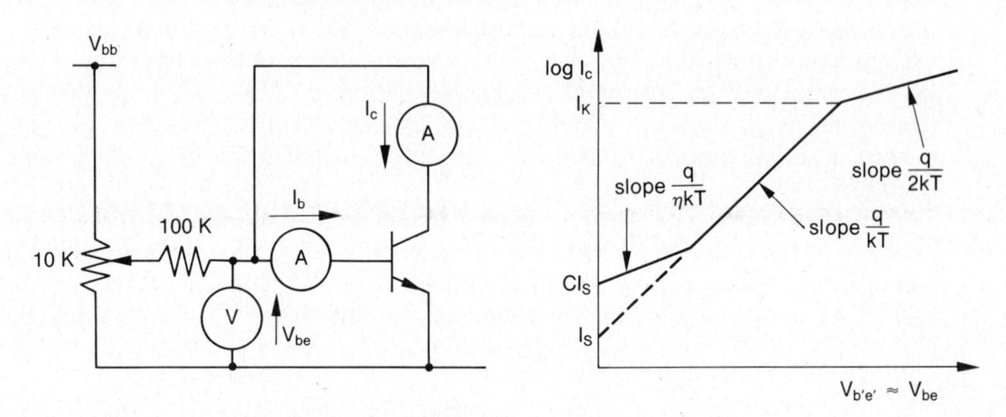

Figure 11.5 Determination of forward (or reverse) saturation current with low and high current parameters

11.4 A chip for determining nMOS process parameters

Apppendix C provides the CIF listing for a chip that allows most process parameters to be determined by direct measurement. The test structures chosen are different from those on multiproject chip starting frames for two reasons. First, additional information can be obtained and, second, the structures are similar to the types required for analog circuits. For example, Van Der Pauw structures for sheet resistance have not been used, but rather simple resistors of varying length and width.

In summary, this analog test structures chip, of dimensions 850λ by 750λ (λ = 2.5 μm), includes:

1. Resistors (polysilicon and n^+ diffusion)—
   (a) 4λ wide:
   - matched pair of seven squares plus two contacts; and
   - a long resistor of 39 squares plus two contacts;

   (b) 2λ wide:
   - matched pair of 14 squares plus two contacts;
   - a long resistor of 78 squares plus two contacts;
2. Snakes—
   (a) metal (858.5 squares);
   (b) butt contact (48 butt contacts plus six metal diffusion contacts in series);

3. Transistors—
   (a) matched pairs of both enhancement and depletion types (width 4λ, length 2λ);
   (b) large enhancement and depletion types (width and length both 40λ);
4. Capacitors—
   (a) depletion implant diode, $110 \times 225\lambda = 24\,750\lambda^2$;
   (b) gate oxide (matched pair), $40 \times 38\lambda = 1520\lambda^2$;
   (c) metal poly (matched pair), $40 \times 120\lambda = 4800\lambda^2$;
   (d) poly to substrate (matched pair), $50 \times 124\lambda = 6200\lambda^2$;
   (e) metal to diffusion (matched pair), $40 \times 120\lambda = 4800\lambda^2$;
   (f) metal substrate (matched pair) $= 9325\lambda^2$
   (which is 3.73 times the standard pad area);
   (g) diffusion to substrate:
   · long diode, 80×20 Area $= 1600\lambda^2$
   Perimeter $= 200\lambda$
   · Square diode, 40×40 Area $= 1600\lambda^2$
   Perimeter $= 160\lambda$

   A second square diode has the glaze removed to compare photo diode performance.

Figure 11.6 shows the metal layer of this analog test structure chip. Pads for DC measurements are at the periphery for bonding while pads for capacity measurements (probed) are internal.

Simple capacitor measurements are made by using an impedance bridge and a probing station, with transistor parameters being determined according to the

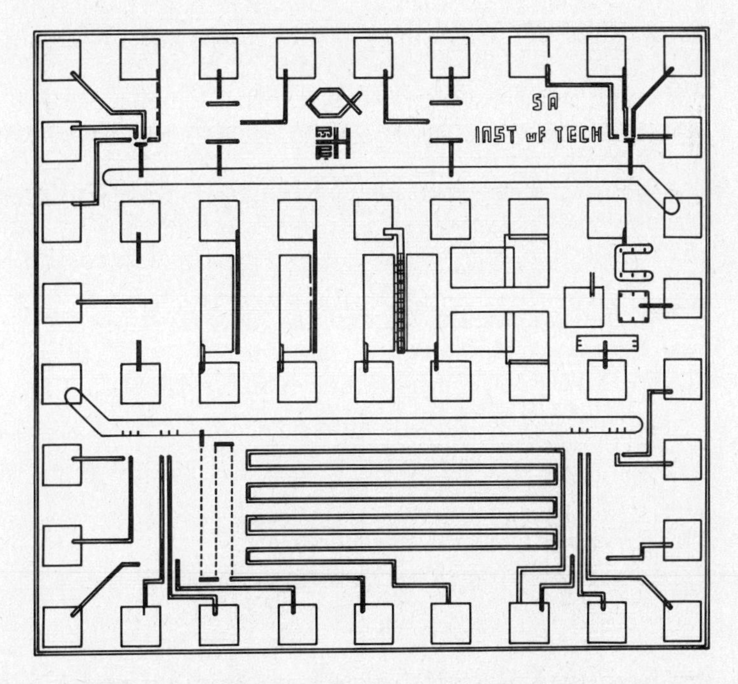

Figure 11.6 Metal/pad layout of analog test structures chip

method described in section 11.2 and snake resistances being measured with a digital multimeter. Resistance measurements should be made using a DC method, otherwise junction and stray capacities to substrate will give incorrect readings.

For resistors of the same width and type, by subtracting resistor values, the actual sheet resistance for resistors of that width will be obtained. Once this is known, the contact resistance can be deduced. Using diffused resistors of differing widths and assuming no photolithographic errors, the side diffusion can be calculated. From these results, the infinite width sheet resistance can be computed and checked with that derived from the starting frame Van Der Pauw structures.

Overall, the chip is extremely versatile and its uses are limited only by imagination.

11.5 Scaling analog circuits

The doubling of complexity of integrated circuits per year (Moore's law) has been achieved partly by increasing chip size and partly by reducing device dimensions. Both are due to the miracles of semiconductor technology and, while physicists may debate just how small a device may be made (and therefore how complex a system can be built on a single chip), there are good reasons why transistors in analog circuits should not be made the smallest possible allowed by a process. For example, if matched devices are required, it would be foolish to make them of minimum geometry as this would be pushing the limits of photolithography. Let us examine what happens when the size of analog circuits are scaled down.

According to the rules of scaling (Paltridge, 1982; Chalterjee, 1980; and Wong and Salama, 1983), a dimensionless scaling factor b is applied to:

1. the linear geometry of the transistor, reducing by the factor b (this includes not only surface geometry but also depth geometry and therefore the oxide thickness and junction depth);
2. the substrate doping, increasing it by a factor b, so that to a first approximation, the threshold voltage scales;
3. All voltages applied to the circuit which reduces those voltages.

What then is the effect of such scaling on a device? Let us consider the drain current and transconductance of a transistor. To simplify things we will assume an ideal model for the linear or nonsaturation region (equation 2.6),

$$I_d = \frac{\mu\epsilon}{D} \times \frac{W}{L} V_{ds} \left((V_{gs} - V_t) - \frac{V_{ds}}{2} \right)$$

and for the saturation region (equation 2.10),

$$I_d = \frac{\mu\epsilon}{2D} \times \frac{W}{L} \left(V_{gs} - V_t \right)^2$$

Scaling these relationships, as outlined above, the results for the nonsaturation region are,

$$\begin{aligned} I_d' &= \frac{\mu\epsilon}{D/b} \times \frac{W/b}{L/b} V_{ds}/b \left(\left(\frac{V_{gs}}{b} - \frac{V_t}{b} \right) - \frac{V_{ds}/b}{2} \right) \\ &= I_d/b \end{aligned} \quad \textbf{(11.13)}$$

and for the saturation region,

$$\begin{aligned} I_d' &= \frac{\mu\epsilon}{2D/b} \times \frac{W/b}{L/b} \left(\frac{V_{gs}}{b} - \frac{V_t}{b} \right)^2 \\ &= I_d/b \end{aligned} \quad \textbf{(11.14)}$$

(′ indicates a scale parameter). Thus the drain current also scales by the scaling factor.

Let us now turn to transconductance, where gm is defined as follows,

$$g_m \underline{\Delta} = \left. \frac{dI_d}{dV_{gs}} \right|_{V_{ds} = \text{constant}} \quad \textbf{(11.15)}$$

By evaluating gm for each region, we find for the nonsaturated region,

$$g_m = \frac{\mu E}{t_{ox}} \times \frac{W}{L} V_{ds} \quad \textbf{(11.16)}$$

and for the saturated region,

$$g_m' = \frac{\mu\epsilon}{D} \times \frac{W}{L}(V_{gs} - V_t) \quad \textbf{(11.17)}$$

When scaling is applied, for the nonsaturated region,

$$\begin{aligned} g_m &= \frac{\mu\epsilon}{D/b} \times \frac{W/b}{L/b} \times V_{ds}/b \\ &= g_m \end{aligned} \quad \textbf{(11.18)}$$

and the saturated region,

$$\begin{aligned} g_m' &= \frac{\mu\epsilon}{D/b} \times \frac{W/b}{L/b} \times (V_{gs}/b - V_t/b) \\ &= g_m \end{aligned} \quad \textbf{(11.19)}$$

Thus, scaling produces no change in the transconductance of the transistor.

Consider now the output resistance (source drain resistance) of the transistor r_d, which is defined as,

$$r_d \triangleq \left.\frac{dV_{ds}}{dI_d}\right|_{V_{gs} = \text{constant}} \tag{11.20}$$

for the transistor in the nonsaturated region as,

$$\left.\frac{dI_d}{dV_{ds}}\right|_{V_{gs} = \text{constant}} = \frac{\mu\epsilon}{D} \times \frac{W}{L}(V_{gs} - V_t - V_{ds})$$

$$= \frac{1}{r_d} \tag{11.21}$$

and the saturation region as,

$$\left.\frac{dI_d}{dV_{ds}}\right|_{V_{gs} = \text{constant}} = 0 \tag{11.22}$$

that is, $r_d = \infty$, which is not unexpected since we are considering an ideal transistor.

Applying scaling to our transistor in the nonsaturation region we obtain the following:

$$\frac{1}{r_d'} = \frac{\mu\epsilon}{D/b} \times \frac{W/b}{L/b}\left(\frac{V_{gs}}{b} - \frac{V_t}{b} - \frac{V_{ds}}{b}\right)$$

$$= \frac{1}{r_d} \tag{11.23}$$

Thus scaling produces no change to the drain source resistance.

Let us now deal with the effect, if any, that scaling has on the simple inverter amplifier that we have used for our pseudoanalog amplifiers?

We have seen earlier that the small signal model of a transistor consists of gate source (C_{gs}) and drain source (C_{ds}) capacities, a drain resistance and a dependent transconductance generator (g_m), as shown in Figure 11.7. We will assume that the depletion load transistor can be modeled by a resistor (r_{d2}) and gate drain capacitor (C_{gd}). Should this amplifier be driving a similar stage, there will also be a load capacitor of C_{gs}. The circuit and the model is shown in Figure 11.7. It can be simplified further, as shown in Figure 11.8.

The mid-frequency gain of the amplifier (which is low enough so that the shunt capacity C has an impedance high enough to be ignored) is,

$$A_v = \frac{V_o}{V_i} = -g_m r_d \tag{11.24}$$

while the cutoff frequency (capacity C has now reduced the gain to half the mid-frequency gain) is,

$$f_c = \frac{1}{2\pi r_d C} \tag{11.25}$$

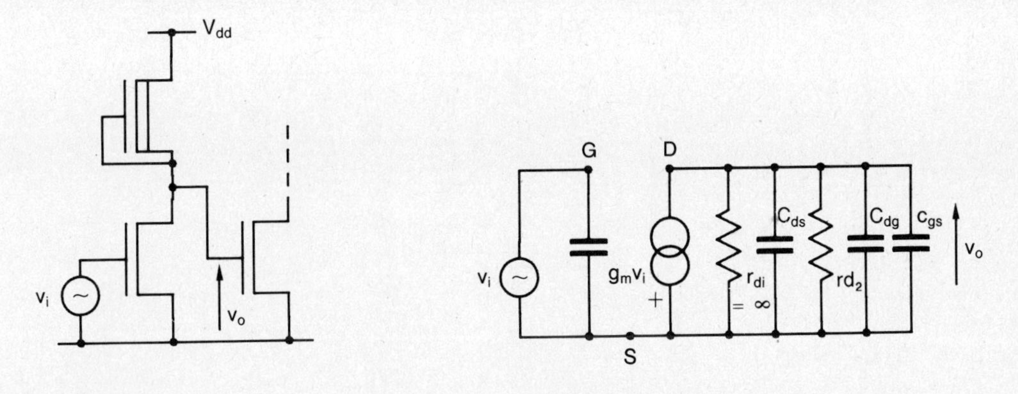

Figure 11.7 Small signal model of an inverting type amplifier

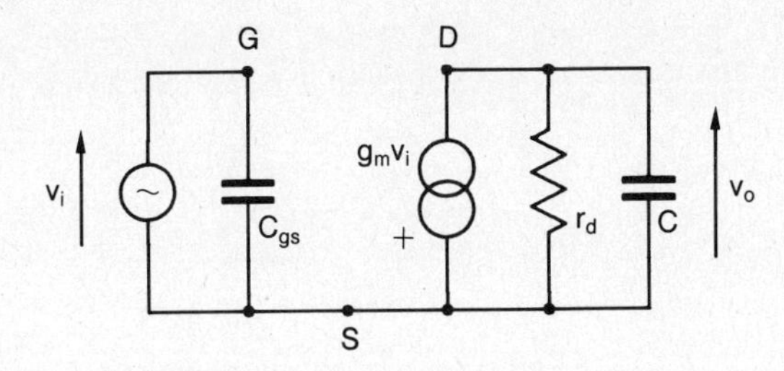

Figure 11.8 Simplified amplifier model

Thus, the magnitude of the gain bandwidth product figure of merit for this amplifier is,

$$|A_v f_c| = \frac{g_m}{2\pi C} \tag{11.26}$$

What is the effect of scaling on these two amplifier parameters? First, we shall deal with voltage gain. By substituting equations 11.19 and 11.21 into 11.24, we obtain the following:

$$A_v' = -g_m r_d = g_m r_d = A_v \tag{11.27}$$

The gain is unchanged. In the case of the gain bandwidth product figure of merit, since capacity is proportional to area divided by thickness (Figure 11.9), then,

$$|A_v' f_c'| = \frac{g_m}{2\pi C'} = \frac{g_m}{2\pi C/b} = b|A_v fc| \tag{11.28}$$

Thus the gain bandwidth product scales directly with b, the scaling factor.

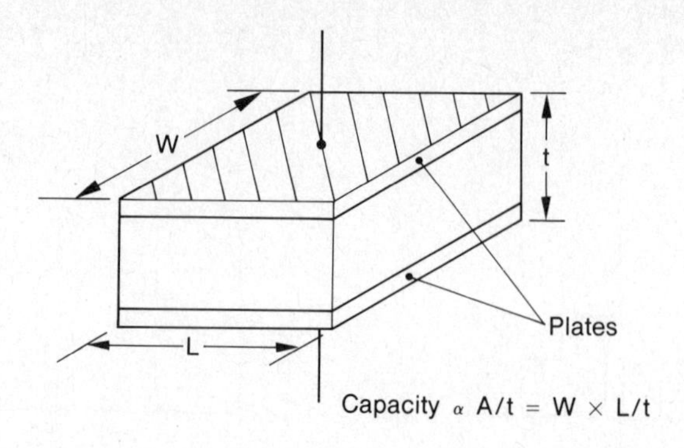

Figure 11.9 Relationship between capacity and device geometry

Now we turn our attention to the effects of scaling on noise. In Chapter 2, it was shown that the noise originating from a transistor can be represented as the noise from an equivalent noise resistor R_{nt}. Reproducing 2.28 and 2.29 we have,

$$V_{gnt}^{\ 2} = 4kTR_{nt}\Delta f \quad \textbf{(11.29)}$$

and,

$$R_{nt} = \frac{2}{3g_m} \quad \textbf{(11.30)}$$

We have already shown in equations 11.18 and 11.19 that the transconductance does not scale, hence R_{nt} does not change. Further, equation 11.28 indicated that the gain bandwidth product scales according to the scaling factor. Consequently, the thermal noise power scales as,

$$(\bar{v}'_{gnt})^2 = 4kTR_{nt}\Delta fb \quad \textbf{(11.31)}$$

or the noise voltage as $b^{\frac{1}{2}}$.

Considering flicker or 1/f noise, from equation 2.30 the following is obtained,

$$\bar{i}^2_{dnF} = \frac{K_1 g_m I_d \Delta f}{C_g f} \quad \textbf{(11.32)}$$

Scaling, we find that,

$$(\bar{i}'_{dnF})^2 = \frac{K_1 g_m I_d / bt_{ox} / b\Delta f}{W/bL/bf\epsilon} = (\bar{i}_{dnF})^2 \quad \textbf{(11.33)}$$

that is, the flicker noise including the spectral density remains constant.

Because flicker noise is a low frequency phenomenon, the frequency f_f, at which the flicker noise spectral density is equal to that of the thermal noise, is important and can be considered as the flicker noise bandwidth. Consequently, using equations 2.27 and 11.32,

$$4kT\left(\frac{2g_m}{3}\right) = \frac{K_1 g_m I_d t_{ox}}{\epsilon W L f_f}$$

or,

$$f_f = \frac{3K_1 I_d t_{ox}}{8kT\epsilon WL} \qquad \textbf{(11.34)}$$

Scaling, we find that $f_f' = f_f$, that is, the noise bandwidth remains unchanged.

Finally, there are the effects of scaling on the dynamic range. For frequencies above f_f, the two factors that limit the dynamic range are the maximum available voltage swing and the noise signal. The first of these is limited by the supply rail, which scales as b, whereas for the second, as we have seen before the thermal noise scales as $b^{\frac{1}{2}}$. Consequently, the dynamic range scales as $b^{-\frac{3}{2}}$.

In summary, it is safe to say that the advantages of scaling an analog amplifier are minimal, for not only does the transconductance and stage gain remain constant, but by keeping the dimensions of devices at the process limits, there are likely to be greater spreads in parameters. While scaling does increase the bandwidth of the stage, it also increases the amount of thermal noise. A summary of the results is given in Table 11.1.

Table 11.1 Effect on amplifier parameters due to scaling

Parameter of amplifier	*Scaling of parameter*
Transconductance gm	1
Stage gain	1
Gain bandwidth product	b
Thermal noise	$b^{\frac{1}{2}}$
Flicker noise	1
Flicker noise bandwidth	1
Dynamic range	$b^{-\frac{3}{2}}$

11.6 Exercises

1 Using the analog test structures chip described in section 11.3, the following average measurements were obtained at 21°C for the diffusion resistors:
- the 10 μm width resistors;
- short resistor = 68.17 ohm; and
- long resistor = 334.24 ohm.

For the 5 μm wide:
- short resistor = 143.17 ohm; and
- long resistor 517.37 ohm.

Calculate the sheet resistances for resistors of 5 μm, 10 μm and infinite width and the resistance of a contact cut at the end of a 10 μm wide resistor. Estimate the amount of side diffusion.

2 Measurements made at 21°C on enhancement transistors which are on the analog test structure chip, in accordance with procedures given in Section 11.2, are as follows:

Test (a)

For $V_{bs} = 0$ V, $W/L = 2$
$V_{ds} = V_{gs} = 2.0$ V, $I_d = 83.3\ \mu A$
For $V_{bs} = -2$ V to restore, $I_d = 83.3\ \mu A$ the new
$V_{ds} = V_{gs} = 2.18$V

Test (b)

For $V_{bs} = 0$ V, $W/L = 1$
$V_{ds} = \frac{1}{2}$ V
With $V_{gs} = 1.0$ V, $I_d = 1.7\ \mu A$
$= 2.0$ V, $= 15.7\ \mu A$
$= 3.0$ V, $= 27.8\ \mu A$

Test (c)

For $V_{bs} = 0$ V, $W/L = 2$
$V_{gs} = 2$ V
With $V_{ds} = 6$ V, $I_d = 124\ \mu A$
$= 2$ V, $= 84\ \mu A$

Assuming the junction depth is 1.0 μm and $2\phi_f$ is 0.65 volts, calculate the remaining important static parameters.

Compare your results with the actual measured characteristics of two devices, $W/L = 2$, as given below. If they do not agree, adjust the value of $2\phi_f$ and recalculate the parameters.

	$V_{bs} = 0$, I_d in μA for V_{ds} =				
V_{gs}	$\frac{1}{2}$ volt	1 volt	2 volt	4 volt	6 volt
1	4.6, 4.4	5.5, 5.3	7.2, 6.8	11.5, 10.7	17.4, 16.2
2	45.2, 44.5	70.6, 69.6	84.9, 83.5	105.8, 103.7	125.3, 121.3
3	78.8, 77.9	151.2, 149.4	210, 207	247, 243	277, 272
4	138, 137	242, 239	352, 347	414, 409	455, 449
5	173, 170	311, 308	489, 482	586, 580	629, 630

11.6 References

Alexander, D. R. et al. (1977), *SPICE 2 MOS Modeling Handbook*, Report BDM/A-77-071-TR, BDM Corporation, New Mexico.

Anderson, H. E. B. and Smith, D. H. (1979), "The Evaluation of MOSFET Model Parameters for Computer Simulation", *Process IREE, Australia*, Vol. 40, No. 5, pp. 148–54.

Chalterjee, P. K. et al. (1980), "The Impact of Scaling Laws on the Choice of n-channel or p-channel MOS VLSI", *IEEE Electron Device Letters*, Vol. EDL-1, No. 10, pp. 220–33.

Getreu, I. (1976), *Modeling the Bipolar Transistor*, Tektronics, Oregon.

Haskard, M. R. (1983), "A Simple Method for Determining SPICE MOS Transistor Model Static Parameters", *Journal of Electrical and Electronic Engineering, Australia,* Vol. 3, No. 3, pp. 232–3.

Hussan, I. H., Camnitz, L. H. and Dally, A. J. (1982), "An Accurate and Simple MOSFET Model for Computer Aided Design", *IEEE Journal of Solid State Circuits*, Vol. SC-17, No. 5, pp. 882–91. IEEE (1981), *IEEE Proc.-I*, "Solid state and Electron Devices", special section on MOSFET Modeling. Vol. 128, No. 6, pp. 218–47.

Kumar, U. (1980), "A Review of Two Dimensional Long Channel MOSFET Modeling", *Microelectronics and Reliability*, Vol. 20, pp. 585–7.

Lui, S. and Nagel, L. W. (1982), "Small-Signal MOSFET Models for Analogue Circuit Design", *IEEE Journal of Solid State Circuits*, Vol. SC-17, No. 6, pp. 983–98.

McCreary, J. (1983), *MOS Device Models—A Tutorial*, Sixth European Conference on Circuit Theory and Design, 4–9 September 1983.

Menckel, G., Borel, J. and Cupcea, N. Z. (1972), "An Accurate Large-Signal MOS Transistor Model for Use in Computer Aided Design", *IEEE Transactions on Electron Devices*, Vol. ED–19, No. 5, pp. 681–90.

Nagel, L. W. (1975), *SPICE 2: A Cumputer Program to Simulate Semiconductor Circuits*, Electronics Research Laboratory, University of California, Berkeley, Memorandum No. ERL-M520.

Paltridge, M. L. (1982), "Scaling Aspects of Analogue nMOS Technology", *Conference Digest*, Microelectronics Conference, Adelaide, 12–14 May 1982, pp. 125–9.

Peng, K. L. and Afromowitz, M. A. (1982), "An Improved Method to Determine MOSFET Channel Length", *IEEE Electron Device Letters*, Vol. EDL-3, No. 12, pp. 360–2.

Schichman, H. and Hodges, D. A. (1968), "Modeling and Simulation of Insulated-Gate Field-Effect Transistor Switching Circuits", *IEEE Journal of Solid State Circuits*, Vol. SC-3, No. 3, pp. 285–9.

Shrivastava, R. and Fitzpatrick, K. (1982), "A Simple Model for the Overlap Capacitance of a VLSI MOS Device", *IEEE Transactions on Electron Devices*, Vol. ED-29, No. 12, pp. 1870–3.

Silburt, A. L., Foss, R. C. and Petrie, W. F. (1984), "An Efficient MOS Transistor Model for Computer-aided Design", *IEEE Transactions on Computer Aided Design*, Vol. CAD-3, No. 1, pp. 104–14.

Simard-Normandin, M. (1983), "Channel Length Dependence of the Body-Factor Effect on nMOS Devices", *IEEE Transactions on Computer Aided Design*, Vol. CAD-2, No. 1, pp. 2–4.

Thoma, M. J. and Westgate, C. R. (1984), "A New AC Measurement Technique to Accurately Determine MOSFET Constants", *IEEE Transactions on Electron Devices*, Vol. ED-31, No. 9, pp. 1113–16.

Vladimirescu, A. (1980), *The Simulation of MOS Integrated Circuits Using SPICE 2*, Electronics Research Laboratory, University of California, Berkeley, Memorandum No. ERL-M.

Ward, D. E. and Doganis, K. (1982), "Optimized Extraction of MOS Model Parameters", *IEEE Transactions on Computer Aided Design*, Vol. CAD-1, No. 4, pp. 163–8.

Wong, S. and Salama, A. T (1983), "Impact of Scaling on MOS Analogue Performance", *IEEE Journal of Solid State Circuits*, Vol. SC-18, No. 1, pp. 106–14.

Appendix A
Layout for selected tiles

KEY

Color encoding	Monochrome mask layout encoding	Layer/feature
Green		$\bar{n}$-diffusion
Red		Polysilicon
Blue		Metal
Yellow		Implant
Black		Contact cut

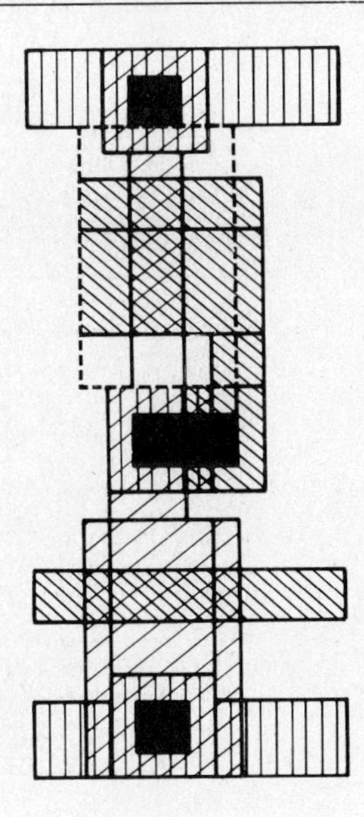

INV9: (Width = W.) A nine to one ratio inverter. Based on the six to one ratio tile.

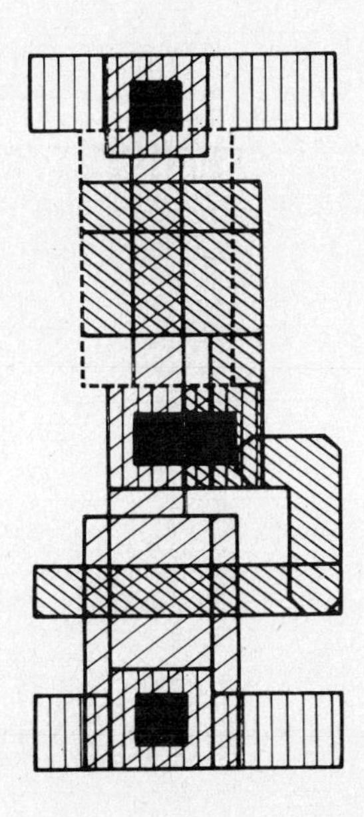

VBIAS9: (Width = W.) A nine to one ratio inverter with its input connected to its output. Used to provide a DC bias voltage.

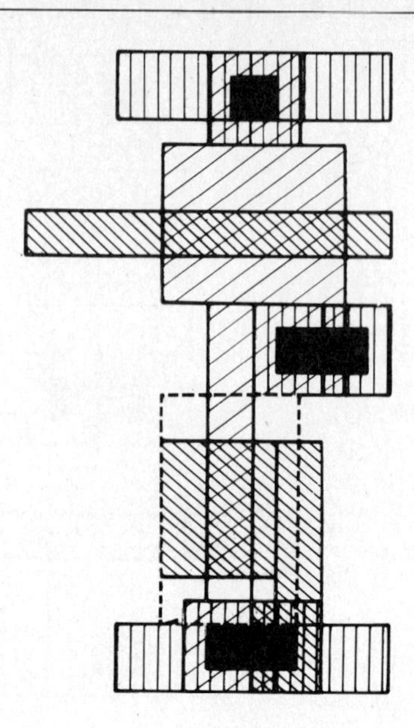

FOLLOW: (Width = W.) A source follower with inbuilt depletion transistor load.

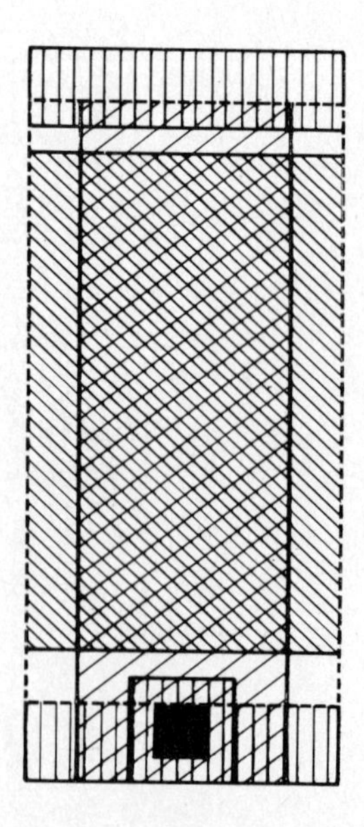

GATECAP: (Width = W.) A polydiffusion overlap capacitor. Produces the highest capacity per unit area but capacitance is very voltage dependent and the diffusion has a large parasitic capacity to substrate.

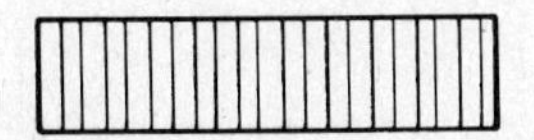

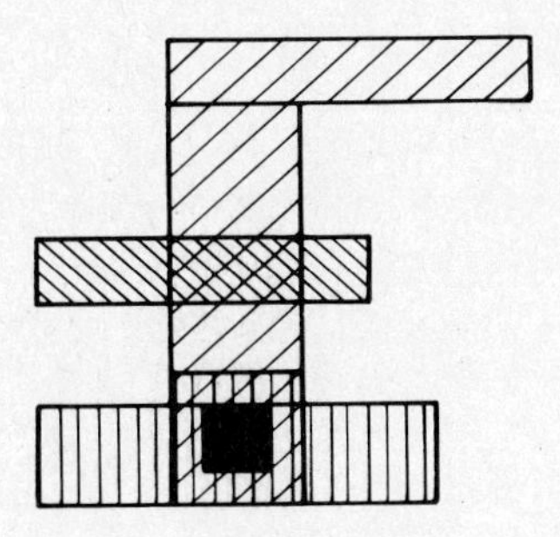

NORADD: (Width = W.) Add on unit for the STATIC BISTABLE tile. Used usually to provide a set input for the bistable.

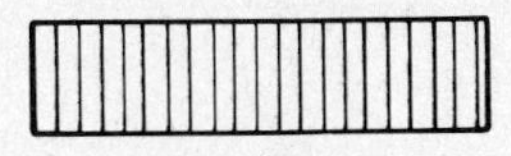

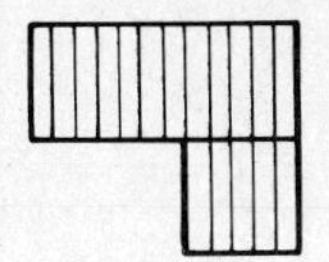

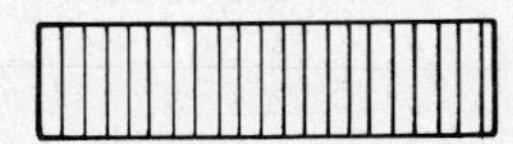

INVOUT: (Width = W.) Used as an overlay for an inverter. It provides the connection between the output and the INVFB tile.

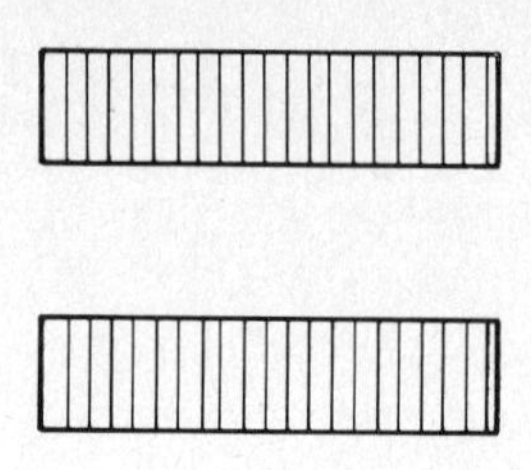

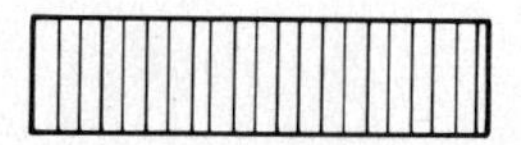

INVFB: (Width = W.) Used as an overlay for an inverter. Connects to the feedback connection point on the BIASCT and SBIASCT tiles.

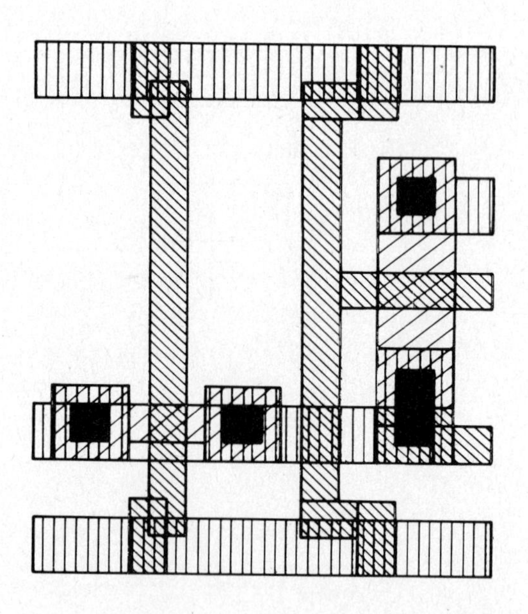

SBIASCT: (Width = W.) A self bias switch with inbuilt overlap capacity compensation.

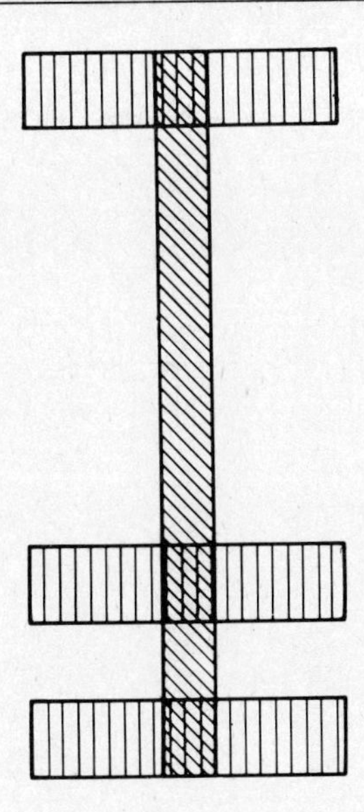

INFEED: (Width = W.) An input coupling tile with clock connection. Always consider the clock coupling caused by the polymetal overlap capacitance.

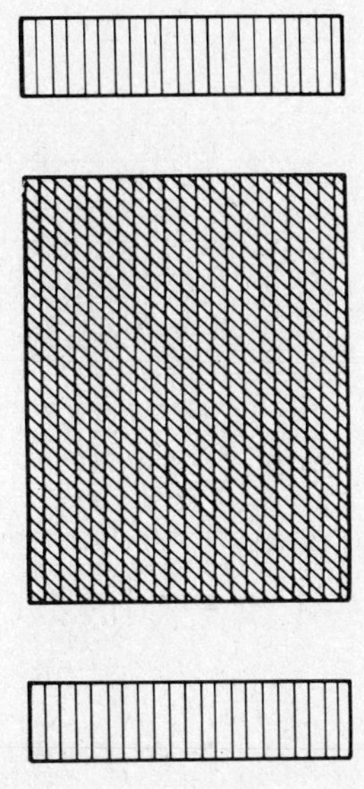

VCAP: A high quality polymetal capacitor. Note that the polysubstrate capacitance is usually about the same as the metalpoly and may determine the connection employed.

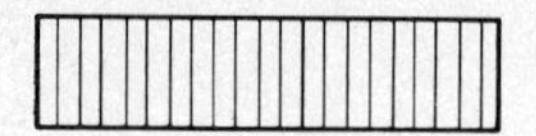

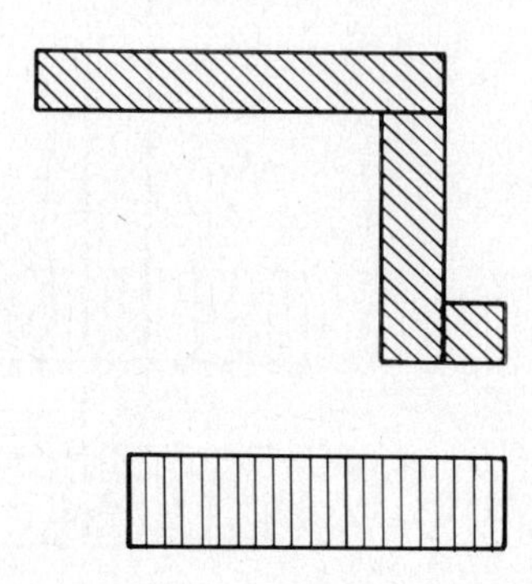

CASCADE: (Width = W.) A polyconnection between one inverter's output and another's input. Note that there is no clock connection provided.

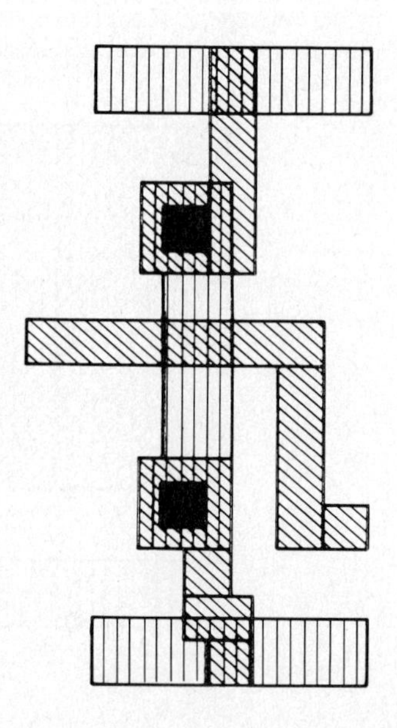

CASCLK: (Width = W.) The same as CASCADE but a clock connection is provided.

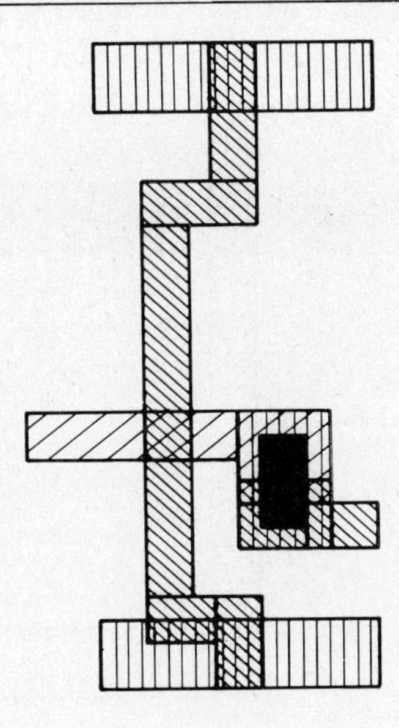

ENTRANS: (Width = W.) An ENHANCEMENT mode series pass transistor.

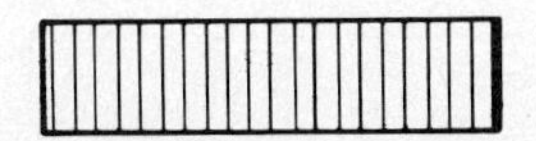

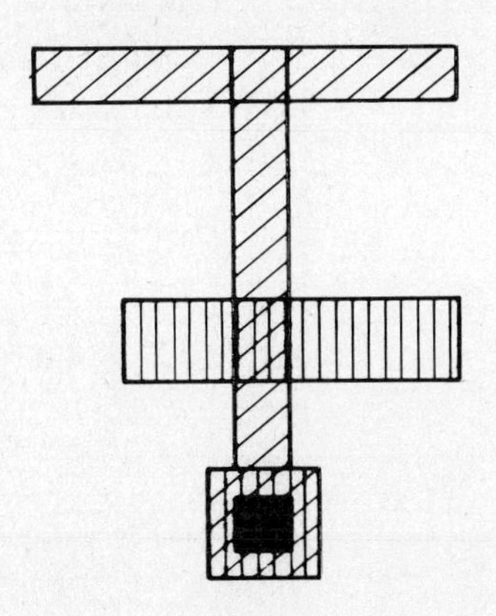

DIFFOUT: (Width = W.) Provides a connection to an inverter's output in diffusion beneath the ground line.

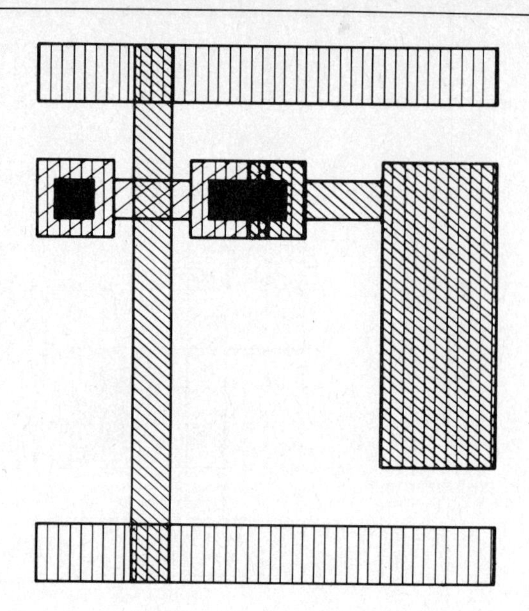

SCAPSWP: (Width = 2*W.) A single switch/capacitor input tile with the switch output connected to the capacitor's poly layer.

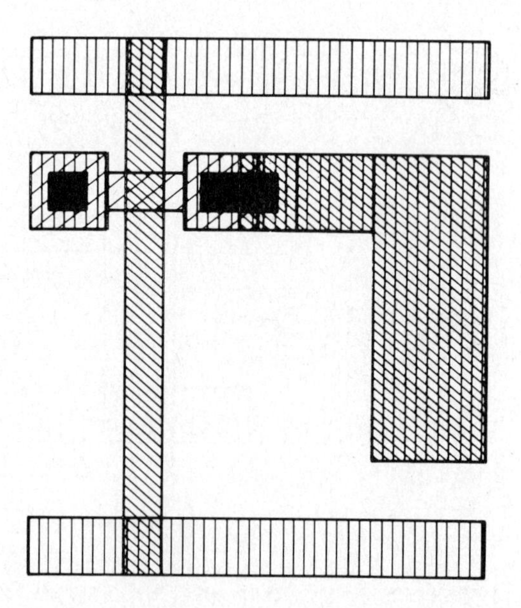

SCAPSWM: (Width = 2*W.) A single switch/capacitor input tile with the switch output connected to the capacitor's metal layer.

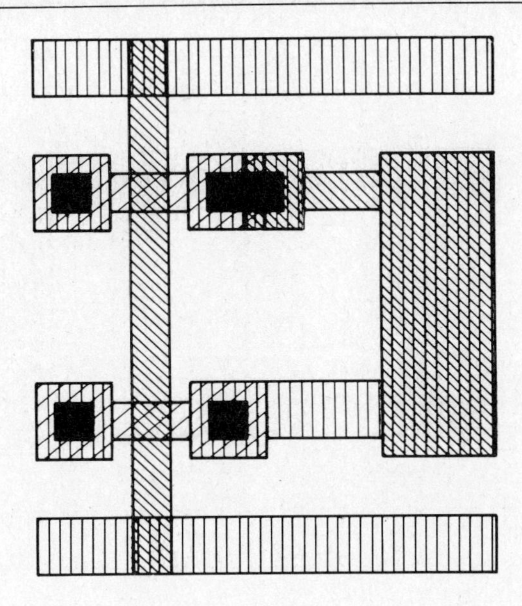

DCAPSW: (Width = 2*W.) A double switch/capacitor input tile. One input connects to the capacitor's poly layer and the other to the metal layer.

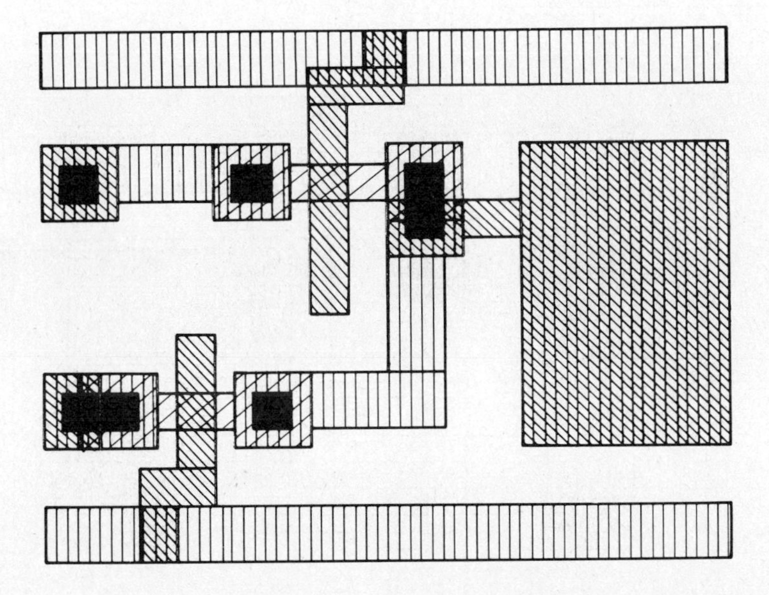

CHSWSER: (Width = 3*W.) A double switch/capacitor input tile that uses a two phase clock signal to connect first one input and then the other to the poly layer of the capacitor. The output is taken from the metal layer of the capacitor and therefore the capacitor is in series with the input. This tile does not connect the clock signal lines from the tile base to top and so results in the minimum clock coupling between the inputs.

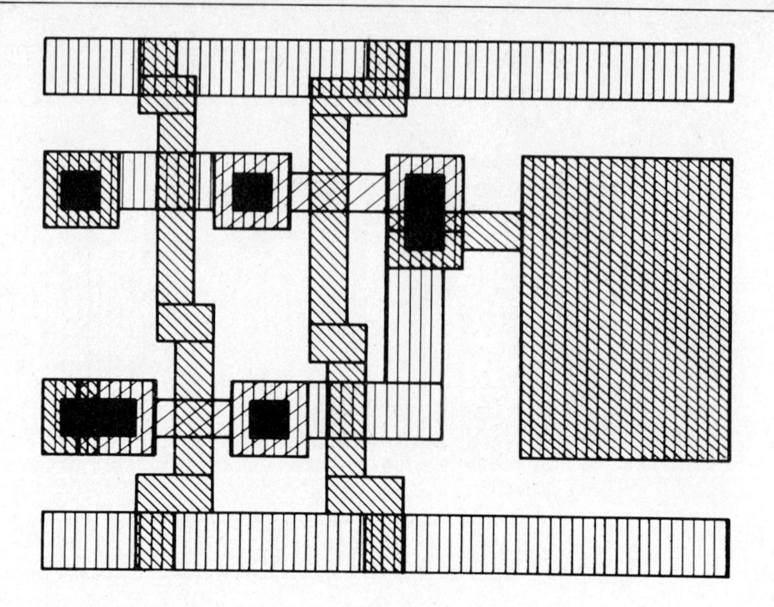

CHSWSERC: (Width = 3*W.) The same as CHSWSER but the clock lines connect from the tile's top to bottom.

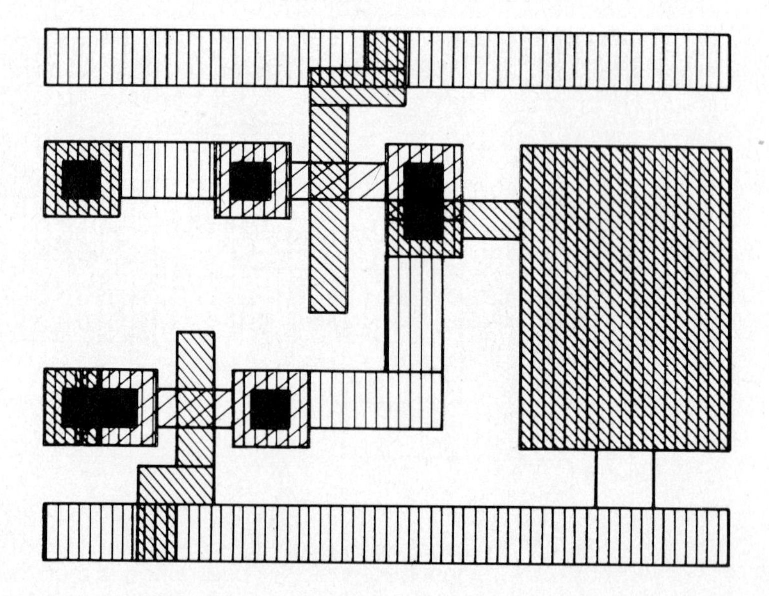

CHSWSH: (Width = 3*W.) The same as CHSWSER but because the capacitor's metal layer is connected to ground the capacitor is in shunt with the input.

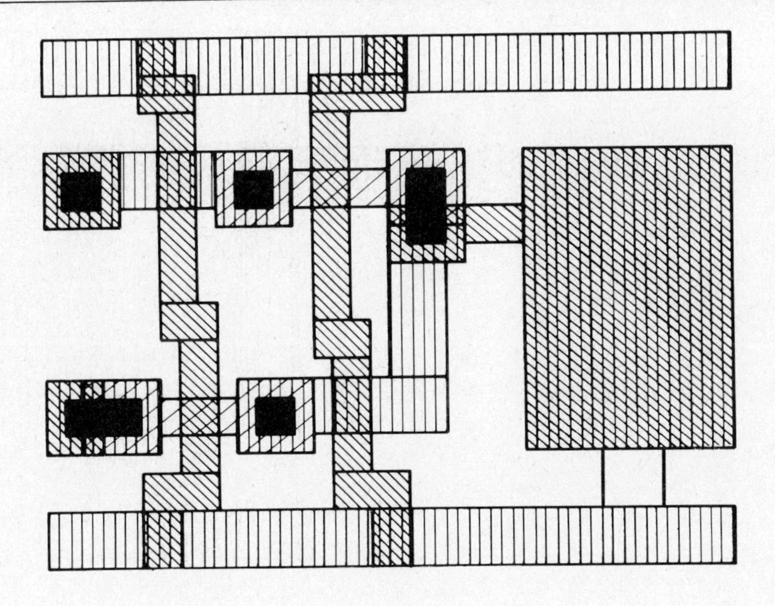

CHSWSHC: (Width = 3*W.) The same as CHSWSH but as for CHSWSERC the clock lines continue from tile top to bottom.

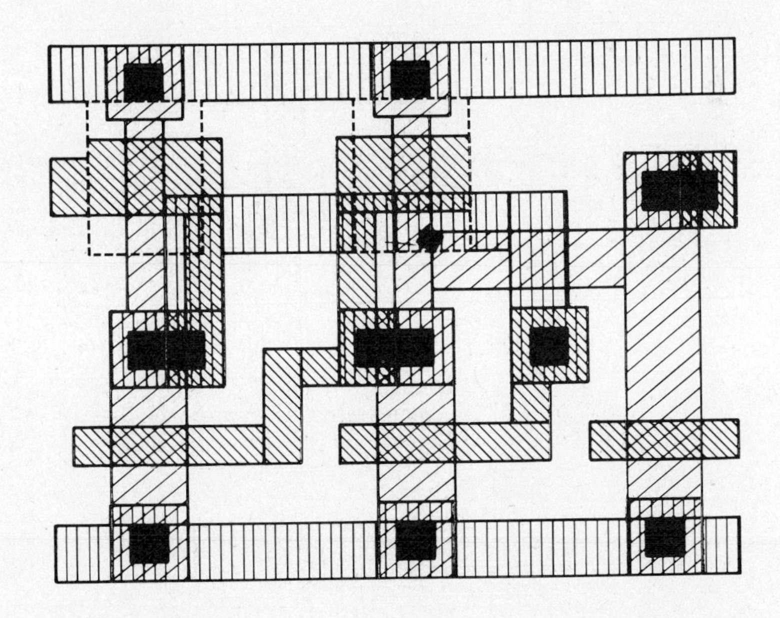

SBIST: (Width = 3*W.) Static bistable tile.

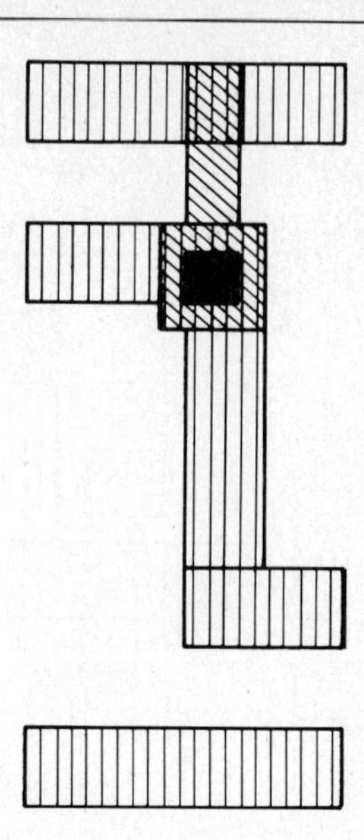

AMPFBHI: (Width = W.) Used to provide an external entry point for feedback to an input. Input connection is at the top of the tile.

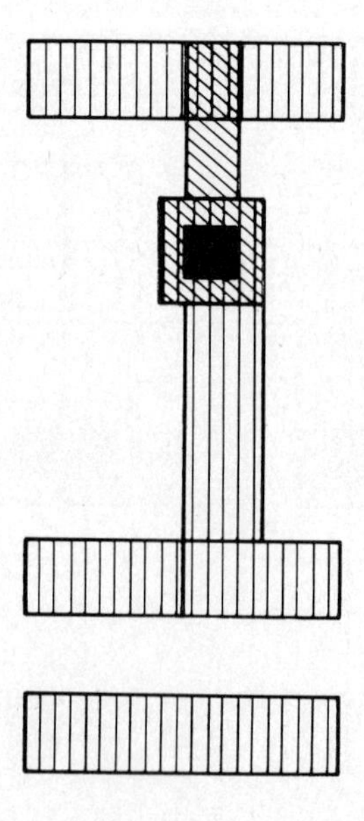

AMPFBLO: (Width = W.) Same as AMPFBHI but input connects at the base of the tile.

Appendix B
Layout for selected pads

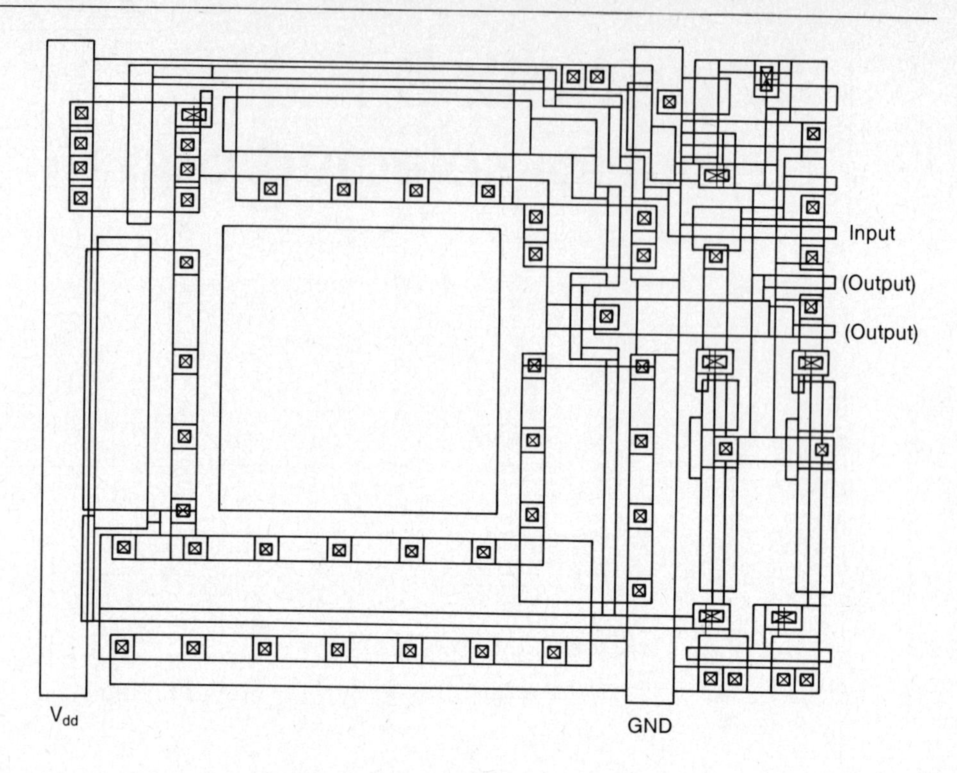

PADOUTV: nMOS analog output pad. Operates from a 5 volt supply.

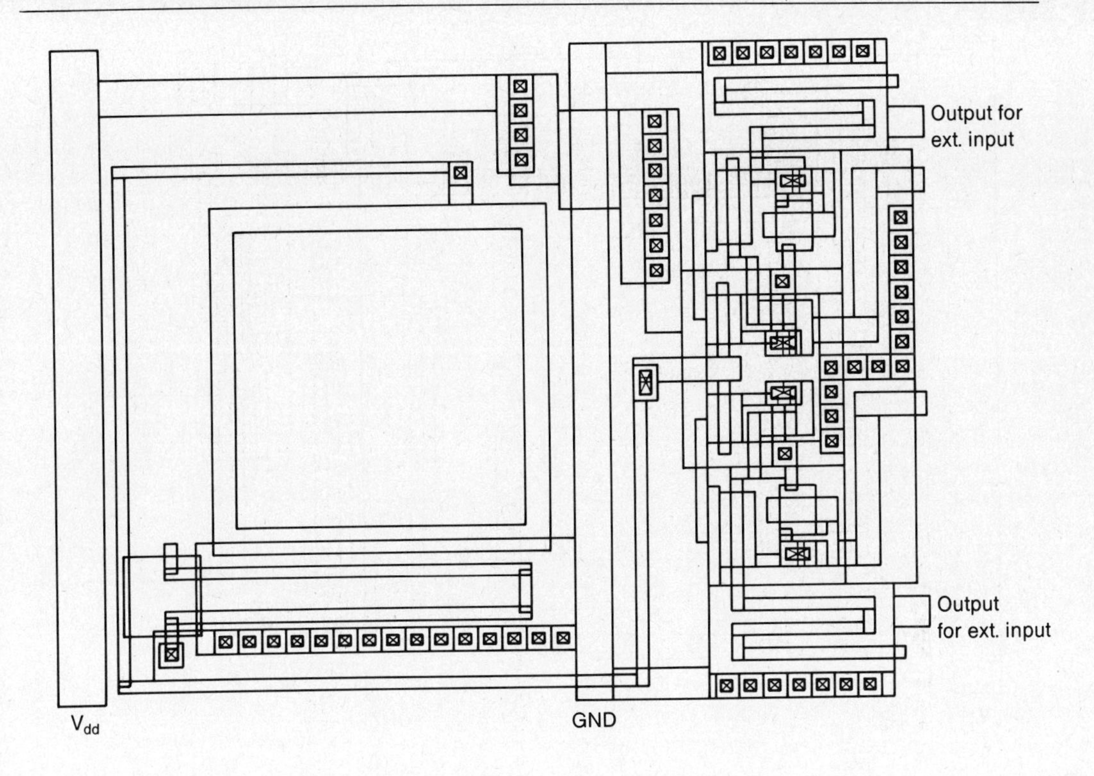

PADCLOCK: nMOS input pad that generates the complement as well as the original signal. Super buffers are included.

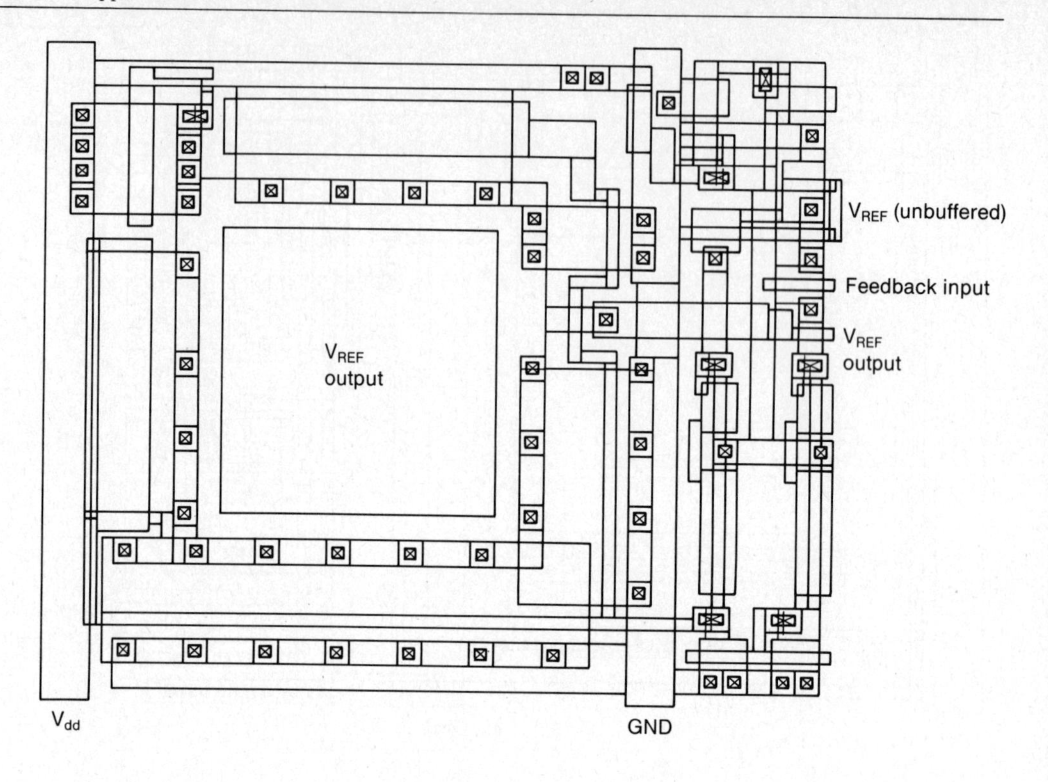

PADREF: nMOS fixed constant voltage reference. Combined with an input pad and an external 100 k ohm potentiometer produces a variable voltage reference.

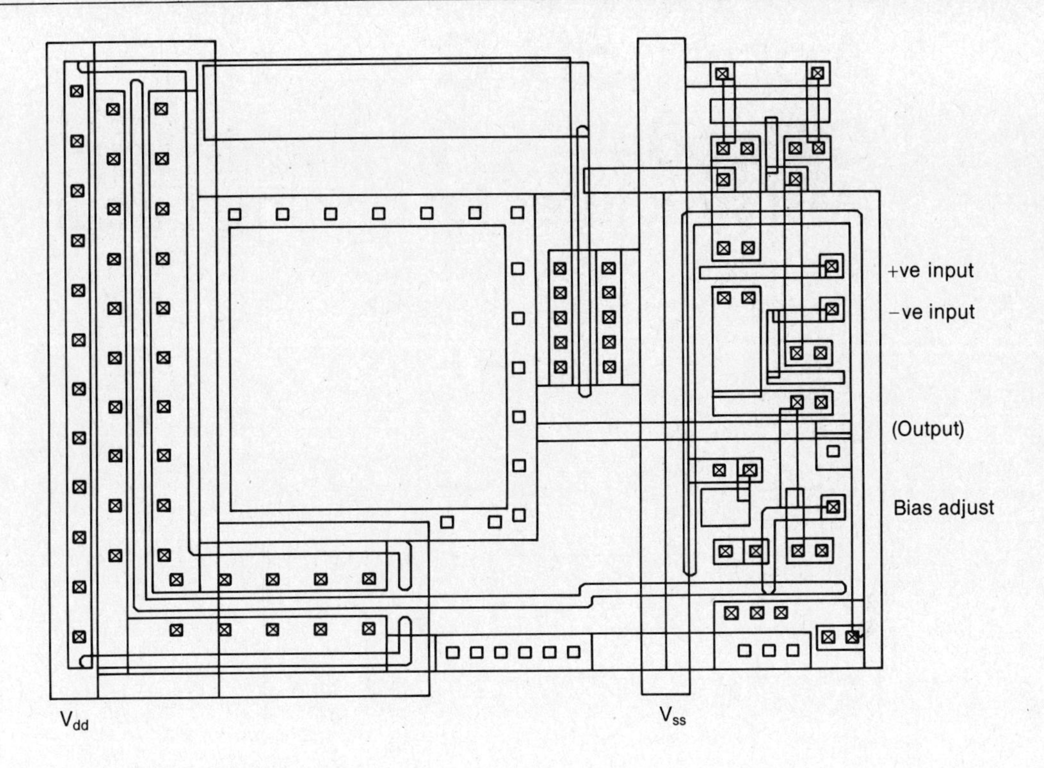

CPADOUTV: CMOS analog output pad. Operates from a 5 volt supply.

Appendix C
CIF code
(test structures for analog circuits)

```
DS 100;
9 PAD;
42 0,0 12500,12500;
        L NM;
                B 12500 12500 6250,6250;
        L NG;
                B 10500 10500 6250,6250;
DF;

DS 102;
9 DIFFCUT;
42 −500,−500 500, 500;
        L ND;
            B 1000 1000 0,0;
        L NC;
            B 500 500 0,0;
        L NM;
            B 1000 1000 0,0;
DF;

DS 101;
9 DIFFUSED-R;
42 0,−41500 18250,0;
        L ND;
                B 1000 9000 500,−4500;
                B 1000 9000 3000,−4500;
                B 1000 41000 5500,−20500;
                B 500 9000 12750,−4500;
                B 500 9000 15250,−4500;
                B 500 41000 17750,−20500;
    C 102 T 500,−500;
    C 102 T 500,−8500;
    C 102 T 3000,−500;
    C 102 T 3000,−8500;
    C 102 T 5500,−500;
    C 102 T 5500,−41000;
    C 102 T 12750,−500;
    C 102 T 12750,−8500;
    C 102 T 15250,−500;
    C 102 T 15250,−8500;
    C 102 T 17750,−500;
    C 102 T 17750,−40500;
DF;

DS 104;
9 POLYCUT;
42 −500,−500 500,500;
    L NP;
        B 1000 1000 0,0;
    L NC;
        B 500 500 0,0;
    L LM;
        B 1000 1000 0,0;
DF;
```

```
DS 103;
9 POLY-R;
42 0,-4100 18250,0;
    L NP;
        B 1000 9000 500,-4500;
        B 1000 9000 3000,-4500;
        B 1000 41000 5500,-20500;
        B 500 9000 12750,-4500;
        B 500 9000 15250,-4500;
        B 500 41000 17750,-20500;
    C 104 T 500,-500;
    C 104 T 500,-8500;
    C 104 T 3000,-500;
    C 104 T 3000,-8500;
    C 104 T 5500,-500;
    C 104 T 5500,-40500;
    C 104 T 12750,-500;
    C 104 T 12750,-8500;
    C 104 T 15250,-500;
    C 104 T 15250,-8500;
    C 104 T 17750,-500;
    C 104 T 17750,-40500;
DF;

DS 106;
9 BUTTCONT;
42 -750,-500 750,500;
    L ND;
        B 1000 1000 -250,0;
    L NP;
        B 750 1000 375,0;
    L NC;
        B 1000 500 0,0;
    L NM;
        B 1500 1000 0,0;
DF;

DS 105;
9 ONE-SNAKE;
42 0,0 1000,42500;
    L ND;
        B 1000 2750 500,1375;
        B 1000 2750 500,41125;
    C 102 T 500,500;
    C 102 T 500,42000;
    L NP;
        B 1000 2500 500,38750;
    L NP;
        B 1000 2500 500,3750;
    L ND;
        B 1000 3000 500,6250;
    C 106 R 0,1 T 500,2500;
```

```
    C 106 R 0,-1 T 500,5000;
    L NP;
        B 1000 2500 500,8750;
    L ND;
        B 1000 3000 500,11250;
    C 106 R 0,1 T 500,7500;
    C 106 R 0,-1 T 500,10000;
    L NP;
        B 1000 2500 500,13750;
    L ND;
        B 1000 3000 500,16250;
    C 106 R 0,1 T 500,12500;
    C 106 R 0,-1 T 500,15000;
    L NP;
        B 1000 2500 500,18750;
    L ND;
        B 1000 3000 500,21250;
    C 106 R 0,1 T 500,17500;
    C 106 R 0,-1 T 500,20000;
    L NP;
        B 1000 2500 500,23750;
    L ND;
        B 1000 3000 500,26250;
    C 106 R 0,1 T 500,22500;
    C 106 R 0,-1 T 500,25000;
    L NP;
        B 1000 2500 500,28750;
    L ND;
        B 1000 3000 500,31250;
    C 106 R 0,1 T 500,27500;
    C 106 R 0,-1 T 500,30000;
    L NP;
        B 1000 2500 500,33750;
    L ND;
        B 1000 3000 500,36250;
    C 106 R 0,1 T 500,32500;
    C 106 R 0,-1 T 500, 35000;
    C 106 R 0,1 T 500,37500;
    C 106 R 0,-1 T 500,40000;
DF;


DS 107;
9 BUT-SNAKE;
42 0,0 11000,42500;
    C 105 T 0,0;
    C 105 T 5000,0;
    C 105 T 10000,0;
    L NM;
        B 6000 1000 3000,500;
        B 6000 1000 8000,42000;
DF;
```

```
DS 108;
9 METAL-SNAKE;
42 -65000,-500 37500,47500;
    L NM;
        B 100000 1000 -15000,17000;
        B 100000 1000 -15000,22000;
        B 100000 1000 -15000,27000;
        B 100000 1000 -15000,32000;
        B 100000 1000 -15000,37000;
        B 100000 1000 -15000,42000;
        B 100000 1000 -15000,47000;
        B 1000 5000 -64500,15000;
        B 1000 5000 -64500,25000;
        B 1000 5000 -64500,35000;
        B 1000 5000 -64500,45000;
        B 1000 5000 34500,20000;
        B 1000 5000 34500,30000;
        B 1000 5000 34500,40000;
        B 1000 6500 500,2750;
        B 18500 1000 -8250,5500;
        B 1000 8500 -17000,9250;
        B 48500 1000 -40750,13000;
        B 1000 11500 23000,5250;
        B 15000 1000 30000,10500;
        B 1000 37500 37000,28750;
        B 3500 1000 35750,47000;
DF;


DS 109;
9 THIN-OXIDE-C;
42 -13750,0 11000, 25500;
    L ND;
        B 10000 21500 5500,12750;
    L NP;
        B 11000 10000 5500,6500;
        B 11000 10000 5500,19000;
        B 1500 1750 5500,875;
        B 1500 1750 5500,24625;
    L NM;
        B 23750 1000 -1875,12750;
    L NI;
        B 11000 22500 5500,12750;
    C 104 T 5500,750;
    C 104 T 5500,24750;
    C 102 T 1250,12750;
    C 102 T 2250,12750;
    C 102 T 3250,12750;
    C 102 T 4250,12750;
    C 102 T 5250,12750;
    C 102 T 6250,12750;
    C 102 T 7250,12750;
```

```
    C 102 T 8250,12750;
    C 102 T 9250,12750;
DF;

DS 110;
9 METAL-POLY-C;
42 0,0 12500,31000;
    L NP;
        B 12500 31000 6250, 15500;
    L NM;
        B 10000 30000 5500,15500;
        B 1000 31000 12000,15500;
    C 104 T 12000,1000;
    C 104 T 12000,4000;
    C 104 T 12000,7000;
    C 104 T 12000,10000;
    C 104 T 12000,13000;
    C 104 T 12000,16000;
    C 104 T 12000,19000;
    C 104 T 12000,22000;
    C 104 T 12000,25000;
    C 104 T 12000,28000;
DF;

DS 111;
9 METAL-DIFF-C;
42 0,0 12500,31000;
    L ND;
        B 12500 31000 6250,15500;
    L NM;
        B 10000 30000 5500,15500;
        B 1000 31000 12000,15500;
    C 102 T 12000,1000;
    C 102 T 12000,4000;
    C 102 T 12000,7000;
    C 102 T 12000,10000;
    C 102 T 12000,13000;
    C 102 T 12000,16000;
    C 102 T 12000,19000;
    C 102 T 12000,22000;
    C 102 T 12000,25000;
    C 102 T 12000,28000;
DF;

DS 112;
9 METAL-C;
42 −18750,−1250 12500,15000;
    L NM;
        B 13750 16250 5625,6875;
        B 20000 12500 −8750,8750;
DF;
```

```
DS 113;
9 L-DIODE;
42 0,0 20000,5000;
    L ND;
        B 20000 5000 10000,2500;
    L NM;
        B 20000 5000 10000,2500;
    C 102 T 1250,1000;
    C 102 T 18750,1000;
    C 102 T 1250,2500;
    C 102 T 18750,2500;
    C 102 T 1250,4000;
    C 102 T 18750,4000;
DF;


DS 114;
9 S-DIODE;
422 0,0 10000,10000;
    L ND;
        B 10000 10000 5000,5000;
    C 102 T 1000,1250;
    C 102 T 1000,8750;
    C 102 T 5000,1250;
    C 102 T 5000,8750;
    C 102 T 9000,1250;
    C 102 T 9000,8750;
DF;


DS 115;
9 S-DIODE-METAL;
41 0,0 10000,10000;
    L NM;
        B 10000 10000 5000,5000;
DF;


DS 116;
9 LIGHT-ST;
42 -1250,-500 11250,10500;
    L NM;
        W 2500 10000,750 0,750 0,9250 10000,9250;
DF;


DS 117;
9 LIGHT-BOX;
42 0,0 10000,10000;
    L ND;
        B 10000 10000 5000,5000;
    C 102 T 5000,1250;
DF;
```

```
DS 118;
9 MATCHED-ENH;
42 -3750,-250 3750,3000;
    L ND;
        B 3500 750 0,375;
        B 1000 3000 -1250,1500;
        B 1000 3000 1250,1500;
    L NP;
        B 2750 500 -1625,1500;
        B 2750 500 1625,1500;
    C 102 T -1250,500;
    C 102 T 1250,500;
    C 102 T -1250,2500;
    C 102 T 1250,2500;
    C 104 T -3250,1500;
    C 104 T 3250,1500;
    L NM;
        B 3000 1000 0,250;
DF;

DS 119;
9 MATCHED-DEP;
42 -2250,750 2250,2250;
    L NI;
        B 2000 1500 -1250,1500;
        B 2000 1500 1250,1500;
DF;

DS 120;
9 LARGE-ENH;
42 -2000,0 10500,13000;
    L ND;
        B 10000 13000 5000,6500;
    L NP;
        B 11000 10000 5000,6500;
        B 2000 1500 -1000,6500;
    L NM;
        B 10000 1000 5000,750;
        B 10000 1000 5000, 12250;
    C 102 T 750,750;
    C 102 T 1250,12250;
    C 102 T 1750,750;
    C 102 T 2250,12250;
    C 102 T 2750,750;
    C 102 T 3250,12250;
    C 102 T 3750,750;
    C 102 T 4250,12250;
    C 102 T 4750,750;
    C 102 T 5250,12250;
    C 102 T 5750,750;
    C 102 T 6250,12250;
    C 102 T 6750,750;
    C 102 T 7250,12250;
```

```
      C 102 T 7750,750;
      C 102 T 8250,12250;
      C 102 T 8750,750;
      C 102 T 9250,12250;
      C 104 T -1250,6500;
DF;

DS 121;
9 LARGE-DEP;
42 -500,1000 10500,12000;
      L NI;
            B 11000 11000 5000,6500;
DF;

DS 122;
9 FISH;
42 -500,-3500 17500,4500;
      L NM;
            W 1000 0,0 4000,4000 10000,4000 17000,-3000;
DF;

DS 123;
9 IDENT;
42 0,0 11000,10000;
      L NM;
            B 1000 4000 500,2000;
            B 4000 1000 3000,1500;
            B 1000 1000 3500,500;
            B 4000 1000 3000,3500;
            B 1000 1000 4500,2500;
            B 1000 2000 500,8000;
            B 1000 2000 2500,8000;
            B 1000 2000 4500,8000;
            B 5000 1000 2500,9500;
            B 11000 1000 5500,5500;
            B 1000 10000 6500,5000;
            B 1000 10000 10500,5000;
DF;

DS 124;
9 SA;
42 0,0 8000,5000;
      L NM;
            W 1000 500,500 2500,500 2500,2500 500,2500 500,4500 2500,4500;
            W 1000 5500,500 5500,4500 7500,4500 7500,500;
            W 1000 5500,1500 7500,1500;
DF;

DS 125;
9 INST;
42 11000,0 51000,5000;
```

```
    L NM;
        W 1000 11500,500 11500,4500;
        W 1000 13500,500 13500,4500 15500,4500 15500,500;
        W 1000 17500,500 19500,500 19500,2500 17500,2500 17500,4500 19500,4500;
        W 1000 21500,4500 23500,4500;
        W 1000 22500,500 22500,4500;
        W 1000 27500,2500 27500,500 29500,500 29500,2500;
        W 1000 27500,2500 29500,2500;
        W 1000 31500,500 31500,4500 32500,4500;
        W 1000 31500,2500 32500,2500;
        W 1000 36500,4500 38500,4500;
        W 1000 37500,500 37500,4500;
        W 1000 42500,500 40500,500 40500,4500 42500,4500;
        W 1000 40500,2500 42500,2500;
        W 1000 46500,500 44500,500 44500,4500 46500,4500;
        W 1000 48500,500 48500,4500;
        W 1000 48500,2500 50500, 2500;
        W 1000 50500,500 50500,4500;
DF;


DS 126;
9 WIRES;
42 -2500,-2500 215000,190000;
    L NM;
(COMMONS);
        W 1000 0,-2000 214500,-2000 214500,189500 -2000,189500 -2000,-2000
                0, -2000;
        W 5000 10000,77500 20000,67500 190000,67500;
        W 5000 22500,145000 192500,145000 202500,135000;
(DIFF R);
        W 1000 25500,58000 10000,58000;
        W 1000 28000,58000 28000,32500 10000,32500;
        W 1000 30500,25500 25500,25500 10000,10000;
        W 1000 37750,58000 37750,20000 32500,20000 32500,10000;
        W 1000 40250,58000 40250,15000 55000,15000 55000,10000;
        W 1000 42750,26000 42750,17500 80000,17500 80000,10000;
(BUT-SNAKE);
        W 1000 50500,62000 50500,66250;
        W 1000 60500,20500 102500,20500 102500,10000;
(POLY-R);
        W 1000 187000,58000 195000,58000 195000,77500 202500,77500;
        W 1000 184500,58000 184500,55000 202500,55000;
        W 1000 182000,26000 192500,26000 192500,30000 202500,30000;
        W 1000 174750,58000 174750,20000 192500,20000 202500,10000;
        W 1000 172250,58000 172250,15000 180000,15000 180000,10000;
        W 1000 169750,26000 169750,17500 160000,17500 160000,10000;
(THIN-OXIDE-C);
        W 1000 30500,94500 30500,85000;
        W 1000 30500,118500 30500,127500;
(METAL-POLY-C);
        W 1000 51000,85000 51000,92500;
        W 1000 62000,90500 62000,127500;
```

```
        W 1000 76000,85000 76000,92500;
        W 1000 87000,90500 87000,127500;
(METAL-DIFF-C);
        W 1000 103500,85000 103500,92500;
        W 2000 115000,91000 115000,127500 110000,127500;
        W 1000 126500,85000 126500,92500;
(L-DIODE);
        W 1000 180500,85000 180500,92500;
(S-DIODE);
        W 1000 192500,104000 202500,104000;
(LIGHT-ST);
        W 1000 187000,121000 187000,127500;
(LIGHT-BOX);
        W 1000 176250,113750 176250,107500;
(MATCHED-ENH);
        W 1000 28000,156500 17500,156500 17500,137000 10000,137000;
        W 1000 30000,157500 30000,160500 10000,160500;
        W 1000 32500,157500 32500,165000 22500,165000 10000,177500;
        W 1000 34500,156500 37000,156500 37000,177500;
        W 1000 30000,155500 32500,155500;
        W 1000 31250,155500 31250,145000;
(LARGE-ENH);
        W 1000 56250,155500 56250,145000;
        W 1000 56250,167500 56250,177500;
        W 1000 63750,161500 77500,161500 77500,177500;
(LARGE-DEP);
        W 1000 123750,161500 110000,161500 110000,177500;
        W 1000 131250,155500 131250,145000;
        W 1000 131250,167500 131250,177500;
(MATCHED-DEP);
        W 1000 185500,156500 182500,156500 182500,172500 162000,172500
               162000,177500;
        W 1000 187500,157500 187500,165000 187000,165000 187000,177500;
        W 1000 190000,157500 190000,165000 202500,177500;
        W 1000 192000,156500 202500,156500;
        W 1000 187500,155500 190000,155500;
        W 1000 188750,155500 188750,145000;
DF;
```

```
(PLACEMENT DEVICES);
    C 100 T 25000,0;
    C 100 T 25000,75000;
    C 100 T 25000,125000;
    C 100 T 25000,175000;
    C 100 T 50000,0;
    C 100 T 50000,75000;
    C 100 T 50000,125000;
    C 100 T 50000,175000;
    C 100 T 75000,0;
    C 100 T 75000,75000;
    C 100 T 75000,125000;
    C 100 T 75000,175000;
    C 100 T 100000,0;
    C 100 T 100000,75000;
    C 100 T 100000,125000;
    C 100 T 100000,175000;
    C 100 T 125000,0;
    C 100 T 125000,75000;
    C 100 T 125000,125000;
    C 100 T 125000,175000;
    C 100 T 150000,0;
    C 100 T 150000,75000;
    C 100 T 150000,125000;
    C 100 T 150000,175000;
    C 100 T 175000,0;
    C 100 T 175000,75000;
    C 100 T 175000,125000;
    C 100 T 175000,175000;
    C 100 T 0,0;
    C 100 T 200000,0;
    C 100 T 0,25000;
    C 100 T 200000,25000;
    C 100 T 0,50000;
    C 100 T 200000,50000;
    C 100 T 0,75000;
    C 100 T 200000,75000;
    C 100 T 0,100000;
    C 100 T 200000,100000;
    C 100 T 0,125000;
    C 100 T 200000,125000;
    C 100 T 0,150000;
    C 100 T 200000,150000;
    C 100 T 0,175000;
    C 100 T 200000,175000;
    C 100 T 167500,97500;
    C 101 T 25000,66500;
    C 103 MX T 187500, 66500;
    C 107 T 50000,20000;
    C 108 T 130000,12500;
    C 118 T 31250,155000;
    C 118 T 188750,155000;
    C 119 T 188750,155000;
```

```
        C 120 MX T 62500,155000;
        C 120 T 125000,155000;
        C 121 T 125000,155000;
        C 109 T 25000,93750;
        C 110 T 50000,90000;
        C 110 T 75000,90000;
        C 111 T 102500,90000;
        C 111 MX T 127500,90000;
        C 112 T 150000,87500;
        C 112 MY T 150000,125000;
        C 114 T 185000,98750;
        C 115 T 185000,98750;
        C 113 T 171250,90000;
        C 114 T 185000,112500;
        C 116 T 185000,112500;
        C 117 T 171250,112500;
        C 126 T 0,0;
        C 122 T 85000,167500;
        C 122 MY T 85000,167500;
        C 123 T 87500,150000;
        C 124 T 157500,165000;
        C 125 T 128500,155000;
E
```

Index